Lecture Notes on Impedance Spectroscopy

Editor

Olfa Kanoun

Chair for Measurement and Sensor Technology
Technische Universität Chemnitz, Chemnitz, Germany

VOLUME 5

CRC Press
Taylor & Francis Group
Boca Raton London New York

CRC Press is an imprint of the
Taylor & Francis Group, an **informa** business

A BALKEMA BOOK

Published by:
CRC Press/Balkema
P.O. Box 447, 2300 AK Leiden, The Netherlands
e-mail: Pub.NL@taylorandfrancis.com
www.crcpress.com – www.taylorandfrancis.com

First issued in paperback 2020

© 2015 by Taylor & Francis Group, LLC
CRC Press/Balkema is an imprint of the Taylor & Francis Group, an informa business

No claim to original U.S. Government works

Typeset by V Publishing Solutions Pvt Ltd., Chennai, India

ISBN 13: 978-0-367-73852-5 (pbk)
ISBN 13: 978-1-138-02754-1 (hbk)

Visit the Taylor & Francis Web site at
http://www.taylorandfrancis.com

and the CRC Press Web site at
http://www.crcpress.com

Lecture Notes on Impedance Spectroscopy, Volume 5 – Kanoun (Ed.)
© 2015 Taylor & Francis Group, London, ISBN 978-1-138-02754-1

Table of contents

Preface

Impedance Spectroscopy is a widely used and interesting measurement method applied in many fields such as electrochemistry, material science, biology and medicine. In spite of the apparently different scientific and application background in these fields, they share the same measurement method in a system identification approach and profit from the possibility to use complex impedance over a wide frequency range and providing interesting opportunities for separating effects, accurate measurements and simultaneous measurements of different and even non-accessible quantities.

For Electrochemical Impedance Spectroscopy (EIS) competency from several fields of science and technology is indispensable. Understanding electrochemical and physical phenomena is necessary for developing suitable models. Suitable measurement procedures should be developed taking into account the specific requirements of the application. Signal processing methods are very important for extracting target information by suitable mathematical methods and algorithms.

The scientific dialogue between specialists of Impedance Spectroscopy, dealing with different fields of science, technology and application, is therefore particularly important to promote the adequate use of this powerful measurement method in both laboratory and in embedded solutions.

The *International Workshop on Impedance Spectroscopy* (IWIS) has been established as a platform for promoting experience exchange and networking in the scientific and industrial field. It was launched already in 2008 with the aim to encourage the sharing of experiences between scientists and to support newcomers dealing with impedance spectroscopy. The workshop has been increasingly gaining acceptance in both scientific and industrial fields and addressing increasingly more fundamentals, but also diverse application fields of impedance spectroscopy. By means of tutorials and special sessions, young scientist get a good overview of different fundamental sciences and technologies helping them to get expertise even in fields which are not in the focus of their previous background.

In 2013 the *Circle of Experts of Impedance Spectroscopy* (CEIS) was founded to promote exchange between experts and together with industry as interest group for promoting Impedance Spectroscopy all over the subfields related to fundamental and applications of Impedance Spectroscopy.

This book is the fifth in the series Lecture Notes on Impedance Spectroscopy which has the aim to widen knowledge of scientists in this field by presenting selected and extended contributions from the International Workshop on Impedance Spectroscopy (IWIS'13). The book reports about new advances and different approaches in dealing with impedance spectroscopy including theory, methods and applications. The book is interesting for researchers and developers in the field of impedance spectroscopy. I thank all contributors for the interesting contributions and the reviewers who supported the decision about publication with their valuable comments.

Prof. Dr.-Ing. Olfa Kanoun

Keynote

Lecture Notes on Impedance Spectroscopy, Volume 5 – Kanoun (Ed.)
© *2015 Taylor & Francis Group, London, ISBN 978-1-138-02754-1*

Potentials of Impedance Spectroscopy for cable fault location

Qinghai Shi & Olfa Kanoun

Chair for Measurement and Sensor Technology, Chemnitz University of Technology, Chemnitz, Germany

ABSTRACT: A new technique is proposed to locate wire faults using Impedance Spectroscopy (IS). The propagation along the cables is analytical, modelled with flexible multi section cascading models utilizing frequency dependent scattering parameters. Therefore, it doesn't have the common numerical method problems that the calculation time is proportional to the wiring length or dependent on cable system complexity and the frequency dependent parameter of the cable are approximated as constant in numerical methods. The transmission line model has the same spectrum as the measured input impedance of Wire Under Test (WUT) so that same practical effects such as skin and proximity effects, signal loss, dispersion and frequency dependent signal propagation can be exactly incorporated. For determination of model parameters an inverse problem should be resolved and Differential Evolution (DE) approach is proposed. The novel method allows locating hard (short and open circuit) and soft (small impedance changes) faults. The results show an excellent reproducibility and a fast processing time which is less than 30 s for a cable with only a single fault.

Keywords: Scattering parameters; the input impedance of the cable; Impedance Spectroscopy; global optimization technique; wire fault location

1   INTRODUCTION

Characterization and location of wire faults is becoming essential because of the higher complexity of wiring systems in electrical systems and power distribution systems. There are several emerging approaches for wire faults location and characterization. The most widely used technique is reflectometry. Generally, a high-frequency signal or a pulse signal is sent down the cable. The reflected signals including information about changes of cable impedance are used to detect wiring faults. Over the last decade, many methods, such as Time Domain Reflectometry (TDR), Frequency Domain Reflectometry (FDR), Ultra Wide Band (UWB) based TDR, Time-Frequency Domain Reflectometry (TFDR) and Spectrum Time Domain Reflectometry (STDR) [1]–[12] were developed. They use different incident signal and signal processing methods. The TDR method has the advantage to identify the type of wire faults. Other methods focus more on improving location accuracy.

Some authors have demonstrated success locating soft faults in a controlled laboratory environment without the impedance changes from mechanical vibration, movement and moisture [13]–[16]. In these literatures the wire is normally fixed on a table or other surface to prevent movement and vibration, and carefully measured with minimal measurement noise.

In [17], the authors have provided the condition of the location of the soft faults (small impedance changes) using TDR, FDR and SSTDR methods: the impedance variation in the environment of the wire system because of the vibration and movement must be smaller than the impedance changes due to the soft faults itself.

Some improved methods [17]–[19] use the baseline method, in which the output signal of the faulty wiring is compared with the output of the healthy wiring, in order to detect and locate soft faults. But it is difficult to obtain a perfect baseline in a realistic environment.

Correlation algorithms have been applied to detect and locate the small discontinuities [5]. An intrinsic limitation of this technique is the attenuation and dispersion of the reflected signal that can limit the maximum distance of wire fault detection and can affect the accuracy of fault location.

Cepstrum [20] and Pecstrum [21] algorithms, which use the system transfer function, have been also applied to detect and locate wire faults. Because the mathematically injected noise of the deconvolution for the transfer function can't be proposed, these techniques were only applied to locate hard faults that cause a large reflection, which has a high SNR in the TDR measurement.

But for all these methods, knowledge of the wire material and geometry is required in advance and the excitation signal and its bandwidth should fulfill high quality requirements. This is because for fault localization, the time of flight is transformed to the location by means of the wave propagation velocity, which is dependent on the wire and the available frequencies of the excitation signal.

Furthermore, soft faults and multiple faults are difficult to detect by the baseline method. Because there are practical difficulties due to noise, multiple reflections, un-known load impedances, mechanical variations and changes of electrical parameters in different wires.

In this paper we propose a novel approach for location and characterization of wire faults by using Impedance Spectroscopy (IS) and a model based approach. The investigation was carried out with coaxial cables, which are normally used in electrical and power systems.

Section 2 explains the model-based approach for localization of cable faults. Section 3 describes the modeling of the transmission line. In Section 4 the parameter extraction using global optimization techniques is described. Section 5 gives the results and analysis. Section 6 gives the conclusion.

2 MODEL BASED APPROACHES FOR WIRE FAULT LOCATION

Some model-based wire fault detection methods are reported including different approaches for modeling and signal processing.

In [22] a frequency-independent RLGC parameter for the transmission lines is modeled for the transmission line and the wave-splitting and Green functions techniques are applied to reconstruct the transmission line. The losses of the conductor and dielectric and the frequency dependence of the parameters of the transmission line are neglected, so that this method is not accurate and is suitable for wire fault location only in a simple transmission lines. The transmitted signal is measured to restore the parameters of the transmission line. For that a measurement from both sides is necessary.

A modified approach in [23] is to apply a transmission line model with Finite-Difference Time-Domain method (FDTD). But this method has some drawbacks.

One of the distinct drawbacks is that the FDTD method is inefficient. For the modeling of small faults the transmission line has to be divided into many small subsections to increase the resolution. Consequently, the simulation time and necessary computational resources increase. This is especially critical for long transmission lines with small faults.

Another problem is that the transmission line is modeled in time domain, so some important frequency-dependent parameters can't be exactly represented. These parameters can only be approximated and idealized in order to simplify the simulation process. These approximations lead to critical errors due to divergence of the parameter extraction. Consequently, the measurement system is not efficient and don't realize a sufficient accuracy. Furthermore, the impulse response is derived from the scattering parameter S11, which is measured in the frequency domain and transformed to the time-domain. This is critical for the resolution and computational time. In [23] only the wire faults with open circuits and some special impedance changes are estimated.

In this study the signal propagation along the cables is analytical modeled with flexible multi section cascading models by frequency-dependent ABCD model. Therefore, it doesn't have the common numerical method problems such as Finite-Difference in Time-Domain

(FDTD) in which the simulation time is proportional to the wiring length and complexity of the transmission line systems. The same occurring effects such as signal loss, dispersion and frequency-dependent signal propagation can be exactly incorporated by the transmission line model. For determination of model parameters an inverse problem should be resolved. For that a robust algorithm is necessary in order to calculate the cable length and load impedance. We propose to use a Genetic Algorithms (GA) approach. The input impedance of the cable system is measured in frequency domain by Impedance Spectroscopy (IS). The ABCD model has the same bandwidth as the measured input impedance of the transmission line system using impedance spectroscopy so that there is a good matching between modeling and measurement data and consequently the optimization technique has a fast convergence and a very good accuracy.

This novel method allows locating hard (short and open circuit) and soft (frays and junctions) faults and measuring the impedance of wire fault and consequently the types of wire fault can be distinguished.

3   TRANSMISSION LINE MODEL

For wire fault detection many requirements for modeling have to be fulfilled. The transmission line should be modeled with high accuracy and efficiency. The frequency-dependent losses of the conductors and dielectric must be considered and the complexity and size of the transmission line models should be reduced. In [24] the ABCD matrix model is applied to simulate the transmission line.

As example, we show in this study the modeling of a coaxial cable. Fig. 1 illustrates the cross section of a simple coaxial cable. The inner conductor has a radius a. The outer conductor (shield) has an inner radius b and thickness Δ. The outer radius of the shield is c. Both of the conductors have the same electrical conductivity σ. The cable interior is filled with a lossy dielectric having a relative permittivity ε_{rel}. The magnetic permeability is assumed to be that of free space μ_0. The relative permittivity is frequency-dependent.

The transmission line model is composed of discrete resistors, inductors, capacitors and conductance. A length l of transmission line can conceptually be divided into an infinite number of increments of length $\Delta l(dl)$ such that per-unit-length resistance R', inductance L', conductance G', and capacitance C' are given. Each of the parameters R', L', and G' is frequency-dependent. For example, R' and L' will change in value due to skin effect and proximity effect. G' will change in value due to frequency-dependent dielectric loss [25]–[27]. From literature [24] we can get these four parameters:

$$R' = real\left[Z'_a(\omega) + Z'_b(\omega)\right] \tag{1}$$

$$L' = imag\left[Z'_a(\omega) + Z'_b(\omega)\right]/\omega + L_{out} \tag{2}$$

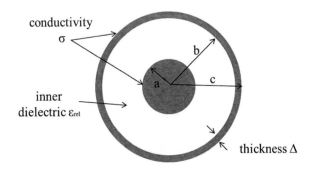

Figure 1.   Cross section a coaxial cable.

$$C' = \frac{2 \cdot \pi \cdot \varepsilon_0 \cdot \varepsilon_{rel}}{\ln(b/a)} \tag{3}$$

$$G' = \frac{2\pi \cdot \varepsilon_0 \cdot \varepsilon''}{\ln(b/a)} = \omega \cdot \tan(\delta) \cdot C' \tag{4}$$

where Z_a and Z_b are the impedances of the inner and outer conductor, respectively, ε'' is the imaginary part of the complex permittivity and the $\tan(\delta)$ is the dielectric loss tangent.

The frequency dependency of the characteristic Impedance of a Transmission Line (TL) Γ can be described by the following equations:

$$\Gamma = \sqrt{\frac{R' + j \cdot \omega \cdot L'}{G' + j \cdot \omega \cdot C'}} \tag{5}$$

The propagation constant of the transmission line with attenuation constant α and phase constant β is:

$$\gamma = \alpha + j \cdot \beta = \sqrt{(R' + j \cdot \omega \cdot L') \cdot (G' + j \cdot \omega \cdot C')} \tag{6}$$

The impedance of a cable with length l and a certain load impedance ZL is:

$$Z_{in}(l) = \Gamma \cdot \frac{Z_l + \Gamma \cdot \tan h(\gamma \cdot l)}{\Gamma + Z_l \cdot \tan h(\gamma \cdot l)} \tag{7}$$

The input impedance of cable with wiring faults can be given as follows [28], [29]:

$$Z_{in} = \frac{A \cdot Z_L + B}{C \cdot Z_L + D} \tag{8}$$

This corresponds directly to the ABCD model.

A typical coaxial cable RG58 C/U is used for the estimation. Fig. 1 shows the cross section of this coaxial cable. The radius of inner conductor is $a = 0.45$ mm, the radius of the dielectric insulation is $b = 1.47$ mm, the outer radius of shield is $c = 1.67$ mm, the dielectric permittivity is $\varepsilon_{rel} = 2.25$, the dielectric loss tangent $\delta = 10^{-4}$ and the conductivity of both conductors is $\sigma = 5.813 \cdot 10^7$ S/m.

4   PARAMETER EXTRACTION

In this study the measured and simulated data of the input impedance of a coaxial cable in the frequency domain are applied to locate wire fault and identify the type of wire fault. The global optimization technique DE is applied to solve the inverse problem because of its good performance of the minimum-seeking and an efficient working with numerically experimental data [30]–[32].

DE is stochastic global optimization algorithm derived from the concept of natural selection and evolution [32]. It uses a special kind of differential operator, which invoked to create new offspring from parent chromosomes, to instead the classical crossover or mutation. DE can be applied to solve a variety of optimization problems that are not well suited for standard optimization algorithms.

The algorithm for parameter extraction is described in Fig. 2. The aim is to calculate the input impedance of the cable system. During the fitting procedure, the calculated input impedance of the cable system is compared with the measured input impedance of the cable system and the input parameter of ABCD model are optimized by DE to find the global

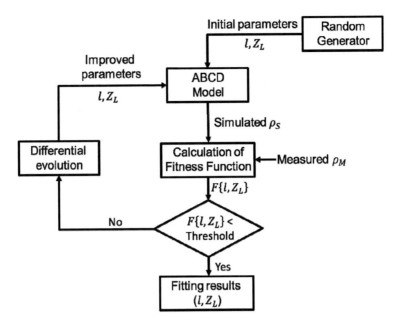

Figure 2. Flowchart of the parameter extraction algorithm.

minimal value of the fitness function. For the impedance measurement a sweep sinusoid signal is applied with equidistant frequencies (N) distributed with a frequency step Δf in the bandwidth BW.

The fitness function of DE in this study is given by (9), where ZM is the measured impedance of the coaxial cable, Z_S is the simulated input impedance of the coaxial cable, df is the frequency step, BW is the bandwidth of the incident sinusoid signals and N is the number of the frequency of the input impedance matrix of the coaxial cable.

$$F\{l, Z_L\} = \sqrt{\frac{1}{N-1} \cdot \sum_{n=1}^{N} \left\{ \frac{\left| |Z_M\{f_n\}| - |Z_S\{f_n, l, Z_L\}| \right|}{|Z_M\{f_n\}|} \right\}} \tag{9}$$

where l is the length between wire fault and input source point, f_n is the frequency of the incident signal, and Z_L is the impedance of wire fault.

For the fitness function a normalization of the difference between simulated and measured data is used in to get a uniform distribution of the deviations at low and high impedance values. With the weighting factor $1/(N-1)$ the results get more independence on the exact number of frequencies.

There are many local minimization values of this fitness function because of its non-linearity. Finding the minimum of a nonlinear function is especially difficult. Typical approaches to solve problems involve either linearizing the problem in a very confined region or restricting the optimization to a small region. Briefly, the constrained parameters are necessary for the quick convergence to find the minimum cost of fitness function. In this study we define the impedance of wire fault changing from 0 Ω to $2 \cdot 10^6$ Ω and the wiring length changing from 0.1 m to 120.1 m. Consequently the minimum cost of $F\{l, Z_L\}$ can be efficiently located. The mutation factor F in this study is 2 and the probability CR is 1. The size of population is 40.

5 RESULTS AND ANALYSIS

As an example, the estimation procedure is applied to simultaneous retrieve the two parameters l and ZL from measured amplitude of the input impedance of a cable system collected from a 5.1- and 100.1 -m-long RG58 C/U coaxial cable with different load impedances.

Fig. 3 shows the number of the period in waveform of the amplitude of the cable's input impedance is proportional to the distance to the fault. The peaks in Fig. 4 are smaller for longer cables because of attenuation and dispersion. The measurement deviation for wire fault location depends on the step frequency of the input signals Δf and the load impedances.

Fig. 5 shows the behavior of the absolute deviation of the wire fault location. The maximal absolute deviation is about 30 cm for different cable lengths. It can be optimized by the set-

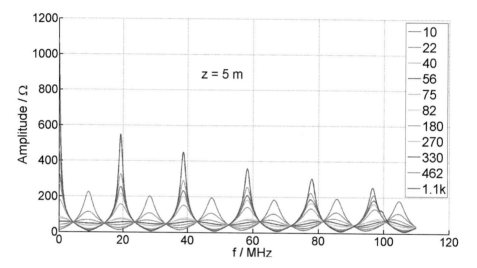

Figure 3.   The input impedance of a 5.1-m-length coaxial cable with different load impedances.

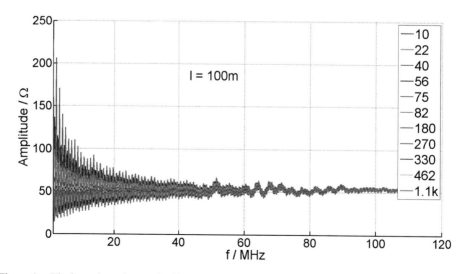

Figure 4.   The input impedance of a 20.1-m-length coaxial cable with different load impedances.

ting new step frequency of the input signals Δf. The larger Δf is, the smaller is the absolute error for wire fault detection. Therefore Fig. 5 shows that the absolute deviation is relatively constant over the lengths estimation from 2.1 m to 100.1 m. This is independent of cable length. If the load impedance of wire fault closely matches the characteristic impedance of the coaxial cable (50 Ω), the measurement system has the maximal deviation (Fig. 6).

The results show an excellent reproducibility and a fast processing time which is less than 30 s for a cable with only a single fault.

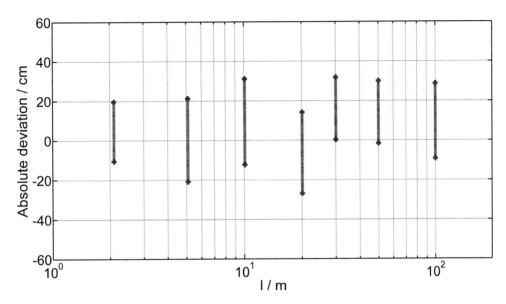

Figure 5.   Reconstruction accuracy of wire fault location of the coaxial cable.

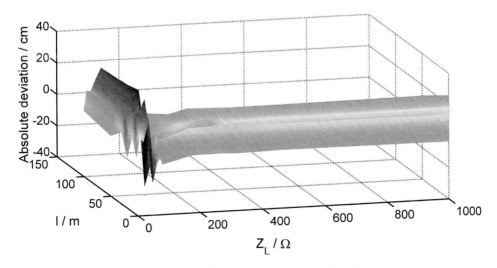

Figure 6.   Reconstruction accuracy of load impedance of the coaxial cable.

6 CONCLUSIONS

This paper has described a forward model for the input impedance of WUT in frequency domain. An ABCD matrix method is used to simulate the transmission lines in frequency domain with frequency dependent parameters. With this method transmission lines can be directly simulated in the frequency domain with realistic results. The required simulation time is independent of the wire length. Because it is an analytical solution and flexible multi section cascading models, so that this method has higher efficiency for the simulation and better accuracy than the time domain transmission line modeling. The novelty in our approach is not to apply RLCG parameters of transmission line or ABCD matrix in frequency domain, but to identify important model parameters and a system structure which are needed to accurately represent the actual hardware measured input impedance of the WUT in the simplest way. Consequently, the transmission line model has the same bandwidth with the measured data so that same practical effects such as signal loss and dispersion and frequency dependent signal propagation can be exactly incorporated. Therefore we can get good matching between simulation and measurement data and consequently the optimization technique has a fast convergence and best accuracy.

The forward model was then combined with a global optimization technique. In this study DE are applied to formulate and solve the problem of optimal fault detection because of its short computational time. Finally, the inverse approach is easily generalized to handle the parameters of forward model and locate the wire fault. Thus, we have presented a truly generalized measurement system applicable to the characterization of IS based wire fault detection.

The reconstruction accuracy of the measurement system depends on the step frequency of the input signals Δf and the load impedance. Clearly, additional study is required to explore the promise of this method including: application this method to estimate faults in different wiring and cables with considering of mechanical variations; estimation of much more complicated wiring systems; identification of the factors affecting the reconstruction accuracy; and a detailed estimation and comparison of different optimization techniques.

REFERENCES

[1] C. Buccella, M. Feliziani, and G. Manzi, "Detection and localization of defects in shielded cables by time-domain measurements with UWB pulse injection and clean algorithm postprocessing," *IEEE Transactions on Electromagnetic Compatibility*, vol. 46, no. 4, pp. 597–605, 2004.

[2] U.K. Chakraborty *et al.*, *Advances in Differential Evolution*. Springer, 2008.

[3] C.-S. Chen and L.E. Roemer, "The application of cepstrum technique in power cable fault detection," in *Acoustics, Speech, and Signal Processing, IEEE International Conference on ICASSP'76.*, IEEE, vol. 1, 1976, pp. 764–767.

[4] C. Furse, Y.C. Chung, C. Lo, *et al.*, "A critical comparison of reflectometry methods for location of wiring faults," *Smart Structures and Systems*, vol. 2, no. 1, pp. 25–46, 2006.

[5] C. Furse, P. Smith, and M. Diamond, "Feasibility of reflectometry for nondestructive evaluation of prestressed concrete anchors," *Sensors Journal, IEEE*, vol. 9, no. 11, pp. 1322–1329, 2009.

[6] K.-H. Gonschorek, *Theoretische Elektrotechnik*. TU Dresden, 2005.

[7] L.A. Griffiths, R. Parakh, C. Furse, *et al.*, "The invisible fray: a critical analysis of the use of reflectometry for fray location," *Sensors Journal, IEEE*, vol. 6, no. 3, pp. 697–706, 2006.

[8] S. Haykins, *Digital Communication*. Willey India, 2010, vol. 11.

[9] J. Lundstedt, S. Strom, and S. He, "Time-domain signal restoration and parameter reconstruction on an LCRG transmission line," in *URSI International Symposium on Signals, Systems, and Electronics, ISSSE'95, Proceedings*, IEEE, 1995, pp. 323–326.

[10] C.R. Paul, *Analysis of Multiconductor Transmission Lines*. John Wiley & Sons, 2008.

[11] K. Price, R.M. Storn, and J.A. Lampinen, *Differential Evolution: A Practical Approach to Global Optimization*. Springer, 2006.

[12] A. Ravindran, G.V. Reklaitis, and K.M. Ragsdell, *Engineering Optimization: Methods and Applications*. John Wiley & Sons, 2006.

[13] S.A. Schelkunoff, "The electromagnetic theory of coaxial transmission lines and cylindrical shields," *Bell System Technical Journal*, vol. 13, no. 4, pp. 532–579, 1934.

[14] Q. Shi and O. Kanoun, "Model-Based Identification of Wire Network Topology," *Journal of the International Measurement Confederation*, vol. 55, pp. 206–211, 2014.

[15] ——, "A New Algorithm for Wire Fault Location Using Time-Domain Reflectometry," *IEEE Sensors Journal*, vol. 14, no. 4, pp. 1171–1178, 2014.

[16] ——, "Automated wire fault location using impedance spectroscopy and Differential Evolution," in *International Instrumentation and Measurement Technology Conference (I2MTC)*, IEEE, 2013, pp. 359–364.

[17] ——, "Wire Fault Location in Coaxial Cables by Impedance Spectroscopy," *IEEE Sensors Journal*, vol. 13, no. 11, pp. 4465–4473, 2013.

[18] ——, "Application of deconvolution for wire fault location using time domain reflectometry," in *IEEE Sensors*, 10/2012.

[19] ——, "Automated wire fault location using impedance spectroscopy and genetic algorithm," in *Sensors, 2012 IEEE*, 10/2012.

[20] ——, "Application of iterative deconvolution for wire fault location via reflectometry," in *Instrumentation & Measurement, Sensor Network and Automation (IMSNA), 2012 International Symposium on*, IEEE, vol. 1, 2012, pp. 102–106.

[21] ——, "A novel method for wire fault location using reflectometry and iterative deconvolution," in *11th International Conference on Signal Processing (ICSP)*, IEEE, vol. 3, 2012, pp. 2139–2143.

[22] ——, "System simulation of network analysis for a lossy cable system," in *9th International Multi-Conference on Systems, Signals and Devices (SSD)*, IEEE, 2012, pp. 1–5.

[23] Q. Shi, U. Troeltzsch, and O. Kanoun, "Detection and localization of cable faults by time and frequency domain measurements," in *7th International Multi-Conference on Systems Signals and Devices (SSD)*, IEEE, 2010, pp. 1–6.

[24] Q. Shi, U. Troltzsch, and O. Kanoun, "Analysis of the Parameters of a Lossy Coaxial Cable for Cable Fault Location," in *8th International Multi-Conference on Systems, Signals and Devices (SSD)*, IEEE, 2011, pp. 1–6.

[25] M.K. Smail, L. Pichon, M. Olivas, *et al.*, "Detection of defects in wiring networks using time domain reflectometry," *IEEE Transactions on Magnetics*, vol. 46, no. 8, pp. 2998–3001, 2010.

[26] P.S. Smith, "Spread spectrum time domain reflectometry," PhD thesis, Utah State University, Logan, Utah, 2003.

[27] E. Song, Y.-J. Shin, P.E. Stone, *et al.*, "Detection and Location of Multiple Wiring Faults via Time-Frequency-Domain Reflectometry," *IEEE Transactions on Electromagnetic Compatibility*, vol. 51, no. 1, pp. 131–138, 2009.

[28] J.P. Steiner and W.L. Weeks, "Time-domain reflectometry for moni-toring cable changes: feasibility study," 02/1990.

[29] L. Van Biesen, J. Renneboog, and A.R. Barel, "High accuracy location of faults on electrical lines using digital signal processing," *IEEE Transactions on Instrumentation and Measurement*, vol. 39, no. 1, pp. 175–179, 1990.

[30] B. Waddoups, C. Furse, and M. Schmidt, "Analysis of reflectometry for detection of chafed aircraft wiring insulation," NASA, 09/2001.

[31] ——, "Analysis of reflectometry for detection of chafed aircraft wiring insulation," PhD thesis, Utah State University, Department of Electrical and Computer Engineering, 2001.

[32] S.C. Wu, "An iterative inversion method for transmission line fault location," PhD thesis, University of Utah, Salt Lake City, 2011.

Biomedical

Lecture Notes on Impedance Spectroscopy, Volume 5 – Kanoun (Ed.)
© 2015 Taylor & Francis Group, London, ISBN 978-1-138-02754-1

Miniaturized wound sensors based on detection of extracellular chromatin

Anna Schröter
Institute of Solid State Electronics, Technische Universität Dresden,
Dresden, Germany

Johannes Wendler, Andreas Nocke & Chokri Cherif
Institute of Textile Machinery and High Performance Material Technology,
Technische Universität Dresden, Dresden, Germany

Angela Rösen-Wolff
Department of Pediatrics, University Hospital Carl Gustav Carus, Dresden, Germany

Gerald Gerlach
Institute of Solid State Electronics, Technische Universität Dresden,
Dresden, Germany

ABSTRACT: Sensors for monitoring wound infections are important to improve care management especially for chronic wounds. As detection parameter the formation of extracellular chromatin was chosen which has characteristic dielectric properties in ionic solvents due to its bound negative charges. Experiments with planar electrodes resulted in a high impedance increase of nearly 450%. The analysis of the relative permittivity revealed a cut-off frequency at 5 kHz. It is shown for the very first time that the changing electrical medium properties during Neutrophil Extracellular Traps (NET) formation are relevant for the occurring dispersion. A textile sensor set-up is proposed to fulfill the requirements of miniaturization and bio-compatibility. With these experimental results it is possible to design a fiber-based sensor based on an impedance detection principle.

Keywords: Wound sensor; Bio-impedance; DNA detection; Fiber-based electrodes

1 INTRODUCTION

Wound infections are a major threat during the treatment of chronic wounds. These wounds have an inhibited healing behavior which indicates the use of hydroactive wound covers, like hydrogels and foams. They keep the wound in a moist environment which enhances the healing process. To overcome side effects like strong pains and massive disturbance of the wound environment it is preferred to change the wound cover as rarely as possible [1]. Hydroactive wound covers can remain longer on chronic wounds, e.g. five to seven days. However, in clinical practice they are changed every two to three days. The reason is to avoid the risk of an unrevealed infection. So, the advantage of a high retention brings along the danger of an infection. A possibility to solve this issue is the development of a wound infection sensor. This would enhance wound treatment.

In previous work we have shown the possibility of detecting an immune defense reaction called the formation of Neutrophil Extracellular Traps (NETs) by impedance spectroscopy on cell cultures [2], [3]. These chromatin structures are released by neutrophils, one kind of white blood cells, during infection. The benefits in detecting chromatin as an infection parameter are

- their fast appearance as neutrophils are in the first front line,
- the massive amount of chromatin released during this chain reaction,
- the inherent reaction with no specificity to the type of pathogen.

Therefore, all infections leading to NET formation can be detected. Due to their origin from the cell core NETs have a DNA backbone with bound histones, which together build the chromatin strand [4]. DNA is strongly negatively charged and has therefore unique dielectric properties. For a suspension of elongated chromatin strands, which appear as large clusters, it is supposable to measure impedance behavior related to the concentration of chromatin.

2   WOUND SENSORS

For an online assessment of the wound status the following requirements have to be fulfilled:

- infection-specific sensitivity,
- miniaturized set-up,
- bio-compatibility,
- disposable or sterilizable system.

There are plenty of approaches which partly fulfill these requirements. Vincenzini et al. and Coyle et al. use optical fibers coated with a functionalization layer [5], [6]. The coating changes its optical properties depending on an input parameter. For pH-sensitivity a pH-responsive hydrogel was used, which changes its swelling behavior. For C-reactive protein detection an antibody-equipped dextran layer was used. The 5 mm thick sensor head is optically connected with a spectrometer. The spectrometer is not miniaturized and has to be placed outside the wound cover. This limits the mobility of the patient. Furthermore, optical fiber wave guides are brittle and tend to break during bending. Cross-sensitivity of wound exudate was not assessed.

Connolly et al. developed an impedimetric sensor for wound moisture [7]. The device measures the moisture level during wound healing with screen-printed electrodes. However, the group reported metrological artifacts with this set-up, e.g. from air enclosures. A direct connection between physiological changes during infection and the measurement results were not part of the investigation.

In contrast to all reported approaches, the goal of this work is to investigate possibilities to measure chromatin in the shape of NETs as the main content of infected wound exudate with an impedimetric sensor. Our sensor set-up evolves a fiber structure based on conducting flexible fibers coated with functional materials. For NET measurement we use an unspecific absorber which allows exudate diffusion. The coated fibers will be wound around each other using textile machining techniques like spinning, braiding or twisting. This allows the easy production of a low-cost miniaturized sensor as a yard good.

3   CHARACTERIZATION OF CHROMATIN IN AN ARTIFICIAL
   WOUND ENVIRONMENT

In a first step we determined the impedimetric behavior of NETs. With some constrains, neutrophils in cell medium model a wound environment. We used commercial interdigitated electrode arrays (Roche xCelligence) equipped with wells for cell culture. This disposable device is meant to have no chemical interaction between the gold electrodes and the standard cultivation medium during impedance measurements. The electrodes were connected via a self-made adapter to an impedance analyzer (ScioSpec IX-3). The experiments were carried out with the standard settings of the device at 12.5 mV amplitude at frequencies from 100 Hz to 1 MHz. First, we measured cell cultures, where we focused on the reaction of the cells. Afterwards the electrical properties only of the medium were investigated by impedance spectroscopy too.

In the cell cultures neutrophils in RPMI-1640 medium were used. They were isolated from human blood by density gradient centrifugation and erythrocyte lysis. We used phorbol 12-myristate 13-acetate to stimulate them by exchanging the medium in the wells. This chemical stimulant mimics the presence of a pathogen which leads to the chromatin release. The control culture did not receive the stimulating agent and was kept unstimulated.

To control whether stimulation was successful we continuously recorded the impedance spectra of the cell cultures for 1 h. As published before the stimulation causes an impedance increase at around 17.5 kHz [3]. Nevertheless, these measurements reflect an unrealistic state where the cells are cultivated on the electrodes. Therefore, we also measured the electrical properties of the medium without growing cells on the electrode to describe the status when the neutrophils arrive and exclude their chromatin in the wound. To determine the electrical medium properties we used the cultivation medium as a reference in a new well. The top layer of cell medium in the cell culture was discarded after a stimulation period of 1 h. The remaining bottom layer containing mostly NETs was suspended in the new well. For the unstimulated control sample this medium contains mostly unstimulated cells. All displayed spectra are representatives for one well.

To evaluate the dielectric properties independently from the electrode geometry, permittivity spectra are of interest, enabling conclusions for other possible electrode settings, e.g. a fiber-based setting. Furthermore, the sensitive frequency range for the permittivity can be extracted. The complex relative permittivity is defined as:

$$\varepsilon_r = (j\omega C_0 \underline{Z}(\omega))^{-1} \qquad (1)$$

where $\omega = 2\pi f$ is the angular frequency, f the excitation frequency at the measuring point, C_0 the air capacity of the electrode well and $\underline{Z}(\omega)$ the complex impedance measured in real and imaginary part at the measuring point. To get rid of parasitic effects (wire resistance and inductance) caused by the measurement set-up, they were determined, modelled in an equivalent circuit and then deducted.

4   IMPEDANCE SPECTRA OF CHROMATIN

Figure 1 shows one representative time course of the impedance change of cell cultures. It is related to the value immediately after stimulation as a reference which is comparable to the unstimulated status before. The unstimulated culture which received cultivation medium without stimulant shows almost no signal change (−0.2%), whereas the stimulated cell culture has already an impedance increase of 26% after 1 h. This increase occurred in all stimulated cultures and is minimally smaller than stated in former publications with 28% to 55% measured after 1 to 4 h [3].

Further experiments were carried out with media from the cultures measured before. Figure 2 shows the spectra of the impedance change caused by the medium alone for media from both, stimulated and unstimulated cells. As stated before, the impedance of unstimulated cultures remain constant after 1 h for all frequencies. However, suspended unstimulated cells influence the impedance of the cultivation medium. For instance, impedance at 153 kHz was increased by ca. 50%. We assume that the high increase particularly at this point is caused by the β-dispersion of the unstimulated neutrophils. The difference between fresh medium and medium with suspended chromatin is even higher (ca. 450%). This reflects the high influence of NET-formation on the medium. Impedance increase by exchange of stimulated medium is also much higher than the impedance increase measured in the stimulated culture itself (26.2%). The frequency of the maximum increase is shifted from it 19 kHz to 49 kHz. We assume that this shift is mainly caused by the detachment of the chromatin from the cell structures and also that the fragile chromatin is cut in smaller fragments. This all can lead to a higher cut-off frequency.

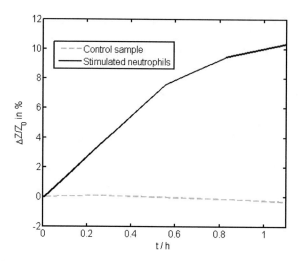

Figure 1. Time course of the deviation of relative impedance change $\Delta Z/Z_0$ for stimulated and unstimulated (control) cells at 20 kHz. Z_0 is the value immediately after stimulation.

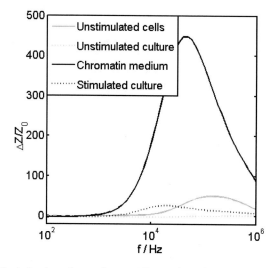

Figure 2. Spectra of relative impedance change $\Delta Z/Z_0$ after stimulation for cultures and for media. Z_0 is for the cultures the value immediately after stimulation and for media the cultivation medium (containing no cells).

Figure 3 depicts the relative permittivity for both, the chromatin-enriched medium and the cell cultivation medium. The cell cultivation medium shows a dispersion at around 150 kHz which is probably caused by the electrode-medium interface. The permittivity spectrum of unstimulated neutrophils is close to that of the cultivation medium and, hence, was left out. A further analysis revealed that a second dispersion is hidden close to 150 Hz supporting the assumption that the impedance increase in Figure 2 for unstimulated cells is caused by a β-dispersion. The medium enriched with chromatin has a further dispersion at around 5 kHz with a decreased real part of the relative permittivity of 18,000. We assume that this dispersion is connected with the NET-formation. The whole spectra are showing a slight negative slope suggesting a further dispersion with a distributed relaxation time. This might be caused by polarization effects of the electrodes.

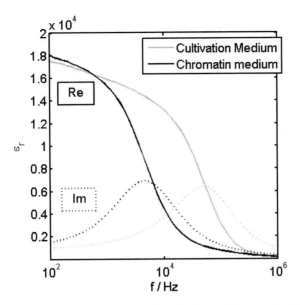

Figure 3. Complex relative permittivity spectrum of cultivation and chromatin-rich medium. The DC conductivity of the medium was about 0.0001 S/m.

5 CONCLUSION AND OUTLOOK

The carried out study confirms the potential of extracellular chromatin as parameter to monitor wound infections. Bound negative charges change the characteristic dielectric properties in ionic solvents. For that reasons, impedance spectroscopy was used as corresponding measurement principle. PMA-stimulated neutrophils caused an impedance increase of around 26% at 20 kHz. This result is in good agreement with published results considering the short stimulation time of 1 h [3]. However, neutrophils in a wound would not settle on the electrodes directly but they would arrive in a suspension. The medium from suspended NET-forming cultures increase the impedance of the medium significantly more than the suspension of unstimulated cells. Also their maxima are located in a different frequency region. This finding supports the assumption that stimulated and unstimulated neutrophils are distinguishable by measuring impedance with a proper excitation frequency. When in the wound environment another medium with different electrical properties can be found, then this frequency could alter. Furthermore, the relative permittivity was determined to get a parameter which is independent from the geometric set-up of the used electrodes. The evaluation for chromatin-rich medium revealed a dispersion at a frequency of 5 kHz. This is of particular importance for the usage of textile electrodes as sensor element. Such a textile set-up should be matched geometrically with the electrical properties of the medium and with the used excitation circuit. Several excitation frequencies should be used to differentiate between the unstimulated and the stimulated state of neutrophils. Next steps will be (i) design the textile fiber electrodes according to the standard textile machining technology and (ii) the evaluation of prototypes with cells with respect to signal stability and sensitivity regarding chromatin.

ACKNOWLEDGEMENTS

The IGF project 17826 BR of the research alliance Forschungskuratorium Textil e.V., Reinhardtstr. 12-14, 10117 Berlin (Germany) was funded by AiF in the frame of the Program for Facilitation of Industrial Corporative Research (IGF) of the Bundesministerium für Wirtschaft und Technologie based on a resolution of the Deutscher Bundestag.

REFERENCES

[1] *Informationsbroschüre Wirtschaftlichkeit und Gesundheitspolitik: Einsatz von hydroaktiven Wundauflagen Bundesverband Medizintechnologie e. V.*, 2011.

[2] A. Schröter, A. Rösen-Wolff, and G. Gerlach, "Impedance-based detection of extracellular DNA in wounds," in *Journal of Physics: Conference Series*, IOP Publishing, vol. 434, 2013, pp. 012–057.

[3] A. Schröter, A. Walther, K. Fritzsche, *et al.*, "Infection monitoring in wounds," *Procedia Chemistry*, vol. 6, pp. 175–183, 2012.

[4] V. Brinkmann and A. Zychlinsky, "Neutrophil extracellular traps: is immunity the second function of chromatin?" *The Journal of cell biology*, vol. 198, no. 5, pp. 773–783, 2012.

[5] S. Pasche, S. Angeloni, R. Ischer, *et al.*, "Wearable biosensors for monitoring wound healing," *Advances in Science and Technology*, vol. 57, pp. 80–87, 2009.

[6] S. Coyle, K.-T. Lau, N. Moyna, *et al.*, "BIOTEX—Biosensing textiles for personalised healthcare management," *Information Technology in Biomedicine, IEEE Transactions on*, vol. 14, no. 2, pp. 364–370, 2010.

[7] P. Connolly, D. McColl, S. Adams, *et al.*, "Monitoring Moisture Retention at the Wound-Dressing Interface," *Journal of Wound Ostomy & Continence Nursing*, vol. 31, no. 3S, S26–S27, 2004.

Lecture Notes on Impedance Spectroscopy, Volume 5 – Kanoun (Ed.)
© 2015 Taylor & Francis Group, London, ISBN 978-1-138-02754-1

Modeling the dynamics of lung tissue with pulsating blood-flow in pulmonary arteries for bioimpedance simulation

Rauno Gordon
Thomas Johann Seebeck Department of Electronics, Tallinn University of Technology, Tallinn, Estonia

Paul Annus
Eliko Competence Centre in Electronics-, Info- and Communication Technologies, Tallinn, Estonia

ABSTRACT: Extensive research is performed with the bioimpedance method internationally for developing medical applications. Before novel diagnostic methods can be developed in hardware and tested in vivo, research must be made with computer-models of the anatomy and physiology for running computer-simulations of measurement configurations. The modeling performed in this paper is for simulation of bioimpedance signals in organs and the whole body, where lung impedance plays a role. The objective of this work is building a model for estimating the bioimpedance signal from the dynamics occurring in the lung anatomy—the cardiac pulse that is featured in the pulmonary vascular network. 3D model of a 3 cm cube sized lung tissue is created with dynamic arterial network. The arterial system in the lung tissue is modeled with automatic generation of hierarchical vascular network using Constrained Constructive Optimization method. The pulmonary artery pressure pulse is used as the input for the arterial system and the wave propagation speed is modeled as 1.5 m/s in the arterial network. The resulting change in the volume of blood in the pulmonary arterial system is modeled taken into account the distensibility of pulmonary arteries. The propagating wave of blood content in the lungs therefore introduces a specific dynamic pattern in the bioimpedance signal, when measured across the lung.

Keywords: Lung bioimpedance; pulmonary arteries; vascular system; modeling; simulation; pulse wave velocity; arterial tree; constrained constructive optimization

1 INTRODUCTION

Bioimpedance analysis is a method to evaluate the health of a patient with electric signals introduced to the target area on the patient by electrodes and by measuring the passive electric properties of the body [1]. The measured signals can show the health status of the measured area and give an indication of a pathologic condition. Time-dependent signals (electric conductivity and dielectric permittivity) depend a lot on blood content that changes dynamically. In some areas that have larger blood vessels or are closer to the heart and lungs, the blood pulsation due to the heart beat can be evident in the bioimpedance signals as well. The pulsation and the bioimpedance signal that exhibits it can vary due to different conditions that the patient goes through—like increased blood flow in the area (vasodilation) or contracted blood vessels (vasoconstriction). These conditions of the vascular system depend on total patient activity at the time (rest, sleep, athletic performance), temperature conditions (vasculature near the skin can dilate or constrict to regulate body temperature) as well as specific conditions that relate to the health of the body and bodypart/organ measured.

For developing medical applications extensive research is performed in this field internationally. Before novel diagnostic methods can be developed in hardware and tested *in vivo*, research starts out with computer-models of the anatomy and physiology that are used to run computer-simulations of potential measurement configurations (testing in *silico*).

The modeling in this paper is meant for simulation of bioimpedance signal in organs and whole body models, where lung impedance plays a role. The electric impedance dynamics of the lung is a major contributor in medical diagnostic methods like Electric Impedance Tomography (EIT) and other simpler methods for the basic health parameters: cardiac output, breathing and lung heal-state. We also believe it has contributions to Electrocardiographic signal.

Modeling the vascular system for dynamic impedance signal simulation has been rarely done. This is mostly because of the complexity of the vascular system dynamics and the dynamics of the resulting electric/dielectric properties. The vascular system and its dynamics has been extensively modeled and simulated from the perspective of mechanical parameters—pulse wave dynamics and flow rates. These models are often compared with in vivo or in vitro studies and the results are positive [2], [3].

We focus here on the blood content of the lung which changes during a heartbeat. Besides breathing and air volume change in the lung, the blood content change is the second major contributor for dynamic electric properties. A piece of lung tissue as a cube with 3 cm side length is modeled here that includes a dynamic arterial network which exhibits a changing blood content with every heartbeat. The model dynamics features heart pulse pressure characteristics in the pulmonary artery, the pulse wavevelocity and the distensibility of pulmonary arteries to introduce the blood volume changes of the arteries. Work by Wtorek 2005 [4] addressed the changes in electric bioimpedance of blood vessels with changing blood flow speed and the unisotropy of the bioimpedance. We believe this change in unisotropy is negated in this bulk tissue because the vectors of blood vessels and therefore the vectors of changing blood flows in a bulk tissue sample have statistically random directions and average out. So the dynamics of blood flow would not affect the unisotropy of bulk tissue impedance. Although this would not be the case in muscle tissue, where the majority of blood vessels are directed along muscle fibers.

2 METHODS

The model of the anatomy of the arterial network was obtained with automatic vascular network generating algorithm based on the Constrained Constructive Optimization (CCO) principle [5]. The method works by selecting a random point in the predefined tissue volume and making a first blood vessel segment from entrance point into that point. The method then chooses the next random point, connects this to the existing segment optimally: a location for the cross-section is chosen so that it provides the defined blood flow with minimal vascular system volume. Next point is chosen, a segment is drawn to it from the existing vascular network that again maintains the blood flow to target tissue with minimal volume for vascular system. This generates a random vascular system tree (arterial) that is optimized for volume and provides blood to all tissue domain (a number of random locations). The algorithm was implemented in MATLAB and it was used to generate an arterial network with one starting artery and 2000 end segments, totally amounting to 3999 segments (Figure 1). The diameters of the arteries in the 3 cm sample cube were selected from 2 mm starting segment diameter to around 50 µm for most ending segment diameters. The algorithm ran smoothly until 4000 segments and started to get increasingly slower after that and did not produce the best vascular system beyond that. This limited the tissue volume as well as the starting segment and end segments' diameters.

Dynamics was introduced based on systolic and diastolic pressure 24 mm Hg and 10 mm Hg respectively [6], [7] that were used for the single input artery segment. The pressure diminishes in the pre-capillaries down to average 10 mm Hg with negligible variation from the pulsation [7]. The speed of the pulse wave traveling was taken as 1.5 m/s average over the network [8]. The pulse wave introduces distention in the arteries and the distention of the segments travels with the pulse wave velocity down towards the capillaries. The vessel lumen diameters and their dynamics were modeled based on the pressure dynamics [6], [7] together with distencibility profiles [9]–[11] (Figure 2). The dynamics of the vessel diameters is essentially the change in blood volume in the tissue during the heartbeat and this should also result in electric impedance change that can be simulated with this model.

The blood vessel lumen diameter changes dynamics are shown in Figure 2. This shows the dynamics of the first input segment of the pulmonary artery—the thickest one in the model. All consecutive artery segments are narrower due to branching, the pressure (as well as pressure pulsation) decreases due to flow impedance and distance from the root segment, and the pulse wave is delayed due to distance from the root segment according to a 1.5 m/s speed.

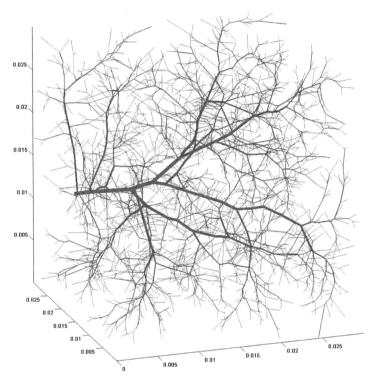

Figure 1.   Automatically generated 3D arterial tree for a 3 cm cube of lung tissue consisting of 3999 total segments (Values on the coordinate axes a given in [m]).

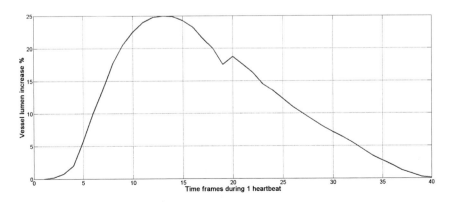

Figure 2.   Pulmonary artery lumen diameter change percentage during the modeled heartbeat.

For the dynamic lung impedance model to be useable in Finite Difference Method or Finite Element Method impedance signal simulations, the dynamic tissue sample model is discretized into volume data. At first 3D data with $35 \times 35 \times 35$ voxel resolution is prepared from each of the 40 time frames. This allows for easy import into MATLAB or COMSOL based calculation. The volume data includes percentage of blood vessels (blood) for each of the $35 \times 35 \times 35 \times 40$ voxels. It can readily be transformed into electric/dielectric properties for each voxel with tissue data available on the internet. But data can also be exported with arbitrary resolution depending on calculation-simulation requirements. The simulations are run separately for each of the 40 time-frames to get full frequency characteristic of impedance measurement across the tissue sample. Finally we can get 40 frequency characteristics—one for each time-frame—and to see a dynamic electrical impedance signal on a certain frequency, we just need to plot the impedance value at the chosen frequency from the 40 time-frames.

3 RESULTS

The result model is a 4D voxel data with 3D lung sample of 3 cm cube size that exhibits dynamics of blood content change resulting from heartbeat. It can readily be imported into MATLAB or COMSOL for example for combining with chest models for simulation of dynamic bioimpedance signals. Incorporating this into dynamic chest-models would give us more precise *in silico* experimentation capabilities for medical diagnostics development.

4 DISCUSSION

The usability for this model in FDM was tested with a 2D version of the same model in previous article [12]. The similarity of the arterial network structure to actual arterial networks have been estimated by the creators of the CCO algorithm [5], but its applicability in final electric impedance simulation is yet to be tested. At the moment the model is not incorporated into a dynamic chest or full body model. The authors are investigating possibilities to add the features of this dynamic model into existing full body dynamic anatomy models for sophisticated and user-friendly modeling methods.

ACKNOWLEDGEMENT

This research was supported by the European Union through the European Regional Development Fund in the frames of the center of excellence in research CEBE, competence center ELIKO, of the Competence Centre program of Enterprise Estonia, National Development Plan for the Implementation of Structural Funds Measure 1.1 project 1.0101.01-0480 and also by Estonian Science Foundation Grant 9394.

REFERENCES

[1] O.G. Martinsen and S. Grimnes, *Bioimpedance and Bioelectricity Basics*. Academic Press, 2011.
[2] K.S. Matthys, J. Alastruey, J. Peiró, *et al.*, "Pulse wave propagation in a model human arterial network: Assessment of 1-D numerical simulations against in vitro measurements," *Journal of biomechanics*, vol. 40, no. 15, pp. 3476–3486, 2007.
[3] J. Alastruey, A.W. Khir, K.S. Matthys, *et al.*, "Pulse wave propagation in a model human arterial network: Assessment of 1-D visco-elastic simulations againstin vitro measurements," *Journal of biomechanics*, vol. 44, no. 12, pp. 2250–2258, 2011.
[4] J. Wtorek and A. Polinski, "The Contribution of Blood-Flow-Induced Conductivity Changes to Measured Impedance," *IEEE Transactions on Biomedical Engineering*, vol. 52, no. 1, pp. 41–49, 2005.

[5] R. Karch, F. Neumann, M. Neumann, *et al.*, "A three-dimensional model for arterial tree representation, generated by constrained constructive optimization," *Computers in Biology and Medicine*, vol. 29, no. 1, pp. 19–38, 1999.

[6] H.K. Hellems, F.W. Haynes, and L. Dexter, "Pulmonary 'Capillary' Pressure in Man," *Journal of Applied Physiology*, vol. 2, no. 1, pp. 24–29, 1949.

[7] W.R. Milnor, C.R. Conti, K.B. Lewis, *et al.*, "Pulmonary Arterial Pulse Wave Velocity and Impedance in Man," *Circulation research*, vol. 25, no. 6, pp. 637–649, 1969.

[8] J.H.S.J. Vander and D.S. Luciano, *Human Physiology*. New York: McGraw-Hill Publishing Company, 1990.

[9] S.H. Bennett, B.W. Goetzman, J. Milstein, *et al.*, "Role of arterial design on pulse wave reflection in a fractal pulmonary network," *Journal of Applied Physiology*, vol. 80, pp. 1033–1033, 1996.

[10] K.L. Karau, R.C. Molthen, A. Dhyani, *et al.*, "Pulmonary arterial morphometry from microfocal X-ray computed tomography," *American Journal of Physiology-Heart and Circulatory Physiology*, vol. 281, no. 6, H2747–H2756, 2001.

[11] S.R. Reuben, "Compliance of the Human Pulmonary Arterial System in Disease," *Circulation research*, vol. 29, no. 1, pp. 40–50, 1971.

[12] R. Gordon and K. Pesti, "System for bioimpedance signal simulation from pulsating blood flow in tissues," *Lecture Notes on Impedance Spectroscopy*, vol. 4, p. 51, 2013.

Lecture Notes on Impedance Spectroscopy, Volume 5 – Kanoun (Ed.)
© *2015 Taylor & Francis Group, London, ISBN 978-1-138-02754-1*

Estimation of Blood Alcohol Content with Bioimpedance Spectroscopy

Mark Ulbrich
Philips Chair for Medical Information Technology (MedIT), RWTH Aachen University, Aachen, Germany

Michael Czaplik
Department of Anaesthesiology, RWTH Aachen University Hospital, Aachen, Germany

Antje Pohl
Philips Chair for Medical Information Technology (MedIT), RWTH Aachen University, Aachen, Germany

Matthias Zink
Department of Cardiology, RWTH Aachen University Hospital, Aachen, Germany

Steffen Leonhardt
Philips Chair for Medical Information Technology (MedIT), RWTH Aachen University, Aachen, Germany

ABSTRACT: Alcohol consumption prior to driving a vehicle is a serious problem and currently the second most common cause of fatal accidents. The gold standard to assess Blood Alcohol Content (BAC) in the evaluation of fitness to drive is the Alcohol Dehydrogenase (ADH) method, using a blood sample obtained from the driver. In daily practice, however, breath alcohol is measured, even though it is less reliable compared with the ADH method. This study examines the feasibility of using bioimpedance as a measure for BAC. For this, Bioimpedance Spectroscopy (BIS) measurements were made in 12 healthy males while they were drinking alcohol, until reaching a BAC of 0.8%. A reference group of 9 healthy individuals drank the same amount of water. In the group drinking alcohol the extracted parameters showed a trend towards higher impedance with increasing alcohol intake; comparison of the extracellular resistance Re at the beginning and the end of the trials, these values showed a significant change (p = 0.042). In contrast, the reference group showed no significant change. In addition, the area under (ROC) curves for Re changes show good detection capabilities for 0.3%, 0.5% and 0.8%. Therefore, under certain conditions, BIS can be used to measure BAC with high specificity and sensitivity.

Keywords: Bioimpedance; spectroscopy; extracellular resistance; body composition; Blood Alcohol Concentration

1 INTRODUCTION

In 2010 in Germany, the second most common cause of accidents (in relation to traffic deaths) was drink-driving, with 23 deaths and 332 seriously injured per 1000 accidents with bodily injury (Figure 1) [1].

Although BAC assessment using the ADH method is the gold standard, in daily practice drink-drivers are generally caught by assessing the alcohol concentration using a breath test, since this is more practicable and feasible. However, due to the unreliability of this method

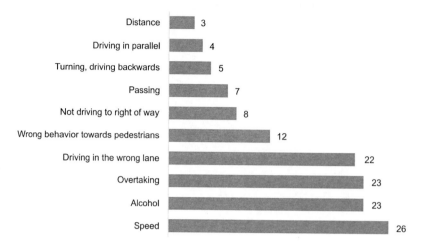

Figure 1. Traffic deaths in Germany per 1000 accidents with bodily injury [1].

in some circumstances, a measurement system is required which delivers reliable BAC values and is easy to use. Other groups have tried to estimate BAC using, for example, transdermal electrochemical sensors [2]. Also, continuous bioimpedance at a fixed frequency has been usedon a finger to assess heart rate variability, which is also assessed using ECG and is not directly related to the measured bioimpedance [3].

This study examines the possibility of the use of spectroscopic bioimpedance measurements to assess BAC.

2   BASICS

2.1   *Alcohol measurement techniques*

Assessment of BAC by taking a blood sample and applying the ADH method is currently the gold standard. ADH is a protein (enzyme) found in high concentrations in liver and the lining of the stomach. It catalyzes the oxidation of ethanol to acetaldehyde according to the following equation:

$$C_2H_5OH + NAD^+ \rightleftharpoons C_2H_4O + NADH + H^+ \tag{1}$$

Here, NAD^+ (Nicotinamide Adenine Dinucleotide) is a coenzyme found in all living cells and $NADH$ is the enzyme in its reduced form. Since NADH is equivalent to the amount of alcohol, the BAC can be assessed by measuring the amount of NADH via photometry at 366 nm or 340 nm [4].

Another possibility to assess the BAC is to use a breath analyzer. In Germany, the only device which is currently accepted in a court process is the Dräger Alcotest 7110 which uses a combination of an electrochemical and an infrared measurement system. The electrochemical sensor contains a membrane covered by an electrolyte which carries the measuring electrode and the counter electrode. The electrolyte and the electrode materials are chosen such that the alcohol to be analyzed at the catalyzing layer of the measurement electrode is electrochemically oxidized. Electrons set free through this reaction correspond to the amount of breath alcohol. The infrared sensor operates at a wave length of 9.5 μm and measures the amount of absorbed energy at this wavelength, which also correlates with the breath alcohol. In addition, gas temperature and the breathing technique are taken into consideration. When combining all of these factors, this device is considered to be very reliable. A breath alcohol concentration of 0.25 mg/l is assumed to equal a BAC of 0.5% [5].

For reasons of economy and simplicity, smaller-sized breath analyzers using an infra-red measurement system only are used for traffic controls. This test with a breath ana-lyzer is authorized and accepted as evidence to detect administrative offences due to alcohol consumption. However, the results of the measurement depend on the distribu-tion of alcohol in the body (during resorption the alcohol concentration is first higher in the lungs and is then redistributed due to blood flow), breathing volume, technique and temperature, eructation and alcohol in the mouth. Therefore, it is not possible to convert a breath alcohol concentration to a BAC with the degree of accuracy that is required [5]. Moreover, breath analyzers can only be used in with the cooperation of the person involved. Therefore, an alternative technology is required to overcome these limitations.

2.2 Bioimpedance

Since alcohol might affect the body composition after intake, with bioimpedance it may be possible to assess changes in body composition caused by alcohol. Application of this technology is an easy and non-invasive way to achieve this aim. Generally, it is possible to measure impedances at one or multiple frequencies. Although many body weight scales are available which assess the body composition using one frequency, a more detailed analysis of body composition can be achieved using BIS at frequencies in a larger frequency range. The range of 1 kHz and 10 MHz (within the so-called β-dispersion region) is the most interesting one for diagnostic purposes, because within this range the physiological and pathophysi-ological processes lead to high impedance changes [6]. However, other factors (such as tem-perature, body position, blood perfusion and metabolic changes) can also alter the measured bioimpedance [7].

BIS is commonly measured between 5 kHz and 1 MHz. A typical frequency locus plot, called Cole plot, is shown in figure 2.

This frequency-dependent semi-circle can be approximated by an electrical equivalent circuit consisting of the extracellular resistance (R_e) parallel to a capacitor (C_m), representing cell membranes, in series with the intracellular resistance (R_i). The resulting impedance is given by equation 2:

$$\underline{Z}(j\omega) = \frac{\underline{u}(j\omega)}{\underline{i}(j\omega)} = \frac{R_e \cdot (R_i + \frac{1}{j\omega C_m})}{R_e + R_i + \frac{1}{j\omega C_m}}$$
$$= \left(\frac{R_e}{R_i + R_e}\right) \cdot \left(R_i + \frac{R_e}{1 + j\omega C_m(R_i + R_e)}\right) \tag{2}$$

For low and high frequencies, the impedance becomes real:

$$\lim_{\omega \to \infty} Z(j\omega) = \frac{R_e \cdot R_i}{R_i + R_e} = R_\infty \tag{3}$$

$$Z|_{\omega=0} = R_e = R_0 \tag{4}$$

Bioimpedance is usually measured with a tetrapolar measurement setup with two elec-trodes injecting a harmless alternating current into the body, and two other electrodes meas-uring the voltage drop over the body part of interest [6]. In practice, this is conducted by standard adhesive Ag/AgCl electrodes. Common electrode positions are whole-body posi-tions (wrist-to-ankle) and hand-to-hand positions (Figure 3).

To conduct whole-body measurements, electrodes are placed on the hand (on the back of the hand by the knuckles, and on the underarm by the wrist) and on the foot (on the back of the foot by the metatarsophalangeal articulations, and on the lower legs by the ankle). For hand-to-hand measurements, both hands are used as explained above.

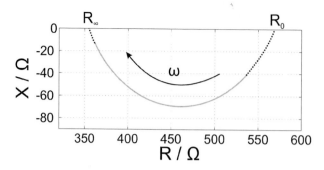

Figure 2. Cole plot for a whole-body measurement. Measured curve with extrapolation (dotted).

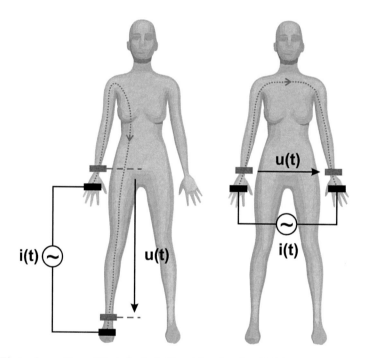

Figure 3. Electrode positions: Whole-body (left) and hand-to-hand (right).

3 METHODS

3.1 *Pre-trial*

A pre-trial was conducted to analyze whether tissues penetrated with alcohol show a measurable effect on impedance measurements. Therefore, three Agar-Agar tissue surrogates were produced with a typical conductivity of human tissue. During the production process, ethanol was inserted in two surrogates in different concentrations, reflecting a realistic amount of alcohol in human tissue. The amount of ethanol per surrogate was calculated using Widmark's equation:

$$A = c \cdot p \cdot r_m \quad \text{with} \quad r_m = 0.715 - 0.00462 \cdot m + 0.0022 \cdot h \tag{5}$$

Here, A is the alcohol mass in mg, c the assumed BAC (max. 2.5‰), p the body mass (75 kg), r_m a reduction factor for men, m again the body mass and h the body height (1.80 m).

The complex impedance of the surrogates was measured between 5 kHz and 1 MHz using a test bench, two metal electrodes, and an LCR meter (Agilent E4980A). A stamp ensured a constant contact pressure between electrodes and surrogate [8].

3.2 Clinical trial

Bioimpedance measurements were conducted with a BIS device (SFB7, ImpediMed, Pinkeba, Australia) covering a frequency range of 5 kHz to 1 MHz. This device has an impedance accuracy of 1% and a phase resolution of 0.1° using adhesive electrodes. To ensure reliable measurements, the body position was not altered and electrode sites were cleaned before attaching the electrodes. For this trial, the whole-body and hand-to-hand bioimpedance is analyzed. Extracellular resistance (R_e), intracellular resistance (R_i) and capacitive effects (C_m) are calculated using the Cole model [6].

Breath alcohol was measured using the Alcotest 7110 Evidential MK III (Dräger Safety, Lübeck, Germany).

After approval by the local Ethics Committee (ID: EK 062/12) and receiving written informed consent, the trial was conducted at the University Hospital (RWTH Aachen, Germany). The study comprised two groups. One group (*ETH*) consisted of 12 individuals drinking alcohol and the second group of 9 individuals drinking the same amount of water (H_2O) (the reference group). All participants were adult males aged 25–40 years. The study protocol was as follows (Figure 4).

First, BIS and BAC reference measurements were made during the first 30 min. Then, 120 ml of vodka (40% abv) was imbibed followed by a resorption phase of 50 min. After this initial phase, all measurements were made and an amount of 60 ml of vodka was imbibed by the subjects followed again by 50 min resorption phase. This cycle was repeated for as long as the subject had not reached a breath alcohol level of 0.4 mg/l, assuming a BAC of 0.8‰. Thus, after the initial baseline measurement, up to 4 measurement cycles were performed. After a subject dropped out of the study, a blood sample of 3 ml was taken as reference. The total trial time was 180 min.

Using the measured impedances, R_e and R_i, the absolute value of the impedance at 100 kHz ($|Z|_{100}$) and the real part of the impedance at 1 MHz ($Re\{Z_{1000,rel}\}$) were extracted. In addition, relative changes of these values were analyzed to include intra-individual changes. A Levene's test was performed to assess the equality of variances of measured data.

To test a significant difference between the two measurements, a two-tailed Student's t-test is the test of choice since the samples are distributed normally. A paired-sample t-test yields a test decision for the null hypothesis that the data in one sample group minus data in the other sample group ($x_1 - x_2$) come from a normal distribution, with mean equal to zero and unknown variance.

$$t = \frac{(\bar{x}_1 - \bar{x}_2) - (\mu_1 - \mu_2)}{\hat{\sigma}_{\bar{x}_1 - \bar{x}_2}} \tag{6}$$

Here, $\bar{x}$ is the sample mean, $\mu_1 - \mu_2$ the theoretical difference of sample means and $\hat{\sigma}$ the standard error of the difference of sample means. The null hypothesis is tested such that the

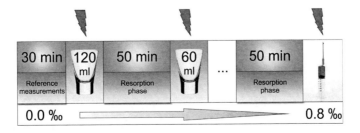

Figure 4. Measurement protocol of the clinical trials.

31

pairwise difference between data x_1 and x_2 has a mean equal to zero. The t-value is used to confirm or reject the null hypothesis. This test was used to verify significant differences with $p < 0.05$ as significance level.

In addition, the Area Under the Curves (AUC) of the Receiver Operating Characteristic (ROC) curves were calculated as a quantitative measure for sensitivity and specificity of BIS to estimate the BAC. In a ROC curve, the True Positive Rate (TPR, sensitivity) is plotted against the False Positive Rate (FPR, 1-specificity). The TPR defines how many correct positive results occur among all positive samples and the FPR defines how many incorrect positive results occur among all negative samples. The AUC is then calculated as follows:

$$AUC = \int_{\infty}^{-\infty} TPR(t) \cdot FPR(t) dt \tag{7}$$

This value (0%–100%) is a measure for the accuracy of a diagnostic measurement method. Classically, the following rating system for AUCs is applied:

- 90–100%: excellent (A)
- 80–90%: good (B)
- 70–80%: fair (C)
- 60–70%: poor (D)
- 50–60%: fail (F)

For statistical analysis, SPSS (IBM SPSS Statistics 19) was used.

4 RESULTS

Figure 5 presents the measurement results of the pre-trial. Here, the complex impedance of the Agar-Agar surrogates without ethanol, and with low and high ethanol concentrations, are depicted.

It can be stated that especially the real part of the impedance of the surrogate increases with increasing ethanol concentration.

Concerning the clinical trial, significant changes for all examined values were analyzed with a focus on measurement point 0 and 3. The extracellular resistance showed the best performance. Figure 6 presents the calculated R_e and measured breath alcohol concentration over the measurement period for one representative subject.

During the whole trial, the resistance increases by 50 Ω. This temporal development is representative for all other subjects, who also show a positive trend for R_e over the measurement

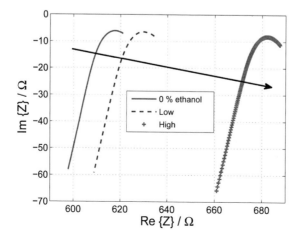

Figure 5.   Measurement results of Agar-Agar surrogates.

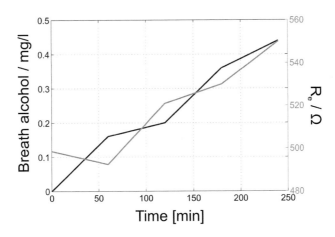

Figure 6. R_e and breath alcohol concentration for one subject over the measurement period.

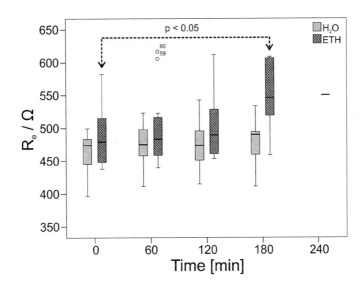

Figure 7. Box plot of R_e values for alcohol (*ETH*) and water (*H₂O*) measurements.

period. This trend can not be seen in the H_2O reference group. Figure 7 plots the R_e values for all subjects in both groups are against measurement time.

This figure shows that the positive trend for R_e can be observed in all subjects, whereas this trend is not present in the reference group. A t-test demonstrated a significant change of R_e between measurement time 0 and 3 (p = 0.042). In the reference group, no significant change was found (p = 0.31).

For all extracted parameters, the AUCs of the ROCs were calculated (Table 1).

The results also reflect the good performance of R_e to compare the sober and drunken state of the subject. Here, the AUC for relative R_e changes in the ETH group is 77.7% while the H_2O group have an AUC of only 58.4%. Nevertheless, the most reliable detection of alcohol levels could not be achieved with the absolute values, but with the relative R_e values. For these, the AUCs of the ROCs were calculated for a BAC detection of 0.3‰, 0.5‰ and 0.8‰. All AUC values show that BAC can be relatively reliably assessed ($AUC_{0.3} = 72.2\%$, $AUC_{0.5} = 88.7\%$, $AUC_{0.8} = 76.9\%$). Figure 8 presents the ROC curve for relative R_e and a BAC of 0.5‰.

Table 1. Area under the curve (measurement point 3) for the ROCs.

Parameter	ETH [%]	H_2O [%]
R_e	77.7	58.4
$|Z|_{100}$	68.9	58.4
$Re\{Z_{1000}\}$	66.6	48.6
R_∞	71.3	45.3
$R_{e,rel}$	85.5	71.6
$|Z|_{100,rel}$	79.4	71.2
$Re\{Z_{1000,rel}\}$	60.1	51.9
$R_{\infty,rel}$	60.1	41.6

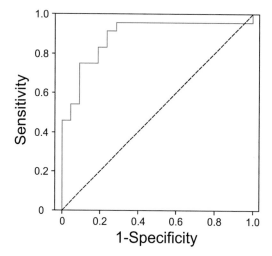

Figure 8.   ROC for relative R_e and 0.5 ‰ BAC.

This graph shows that with an assumed specificity of 70%, a sensitivity of 90% is achieved.

The estimated BACs based on breath alcohol measurement were validated by blood sample analysis at the end of the trial.

5   CONCLUSION

This study examined the feasibility of using BIS to assess the BAC. For this, 12 adult males drank alcohol up to a BAC of 0.8‰ BIS was then measured up to 4 times during the trial to calculate the body composition. In addition, to validate the alcohol measurements, BAC was measured in parallel using a breath alcohol detector as reference.

It was shown that, in all subjects, R_e increases with an increase in BAC whereas consumption of water instead of alcohol does not increase R_e; this demonstrates that this effect is caused by alcohol.

Finally, it is shown that the relative change of R_e is a fair predictor for a BAC of 0.3‰ and 0.8‰ and a good predictor for a BAC of 0.5‰, which is the current limit for driving in Germany. In addition, the ROCs show high sensitivity and specificity. In practice, this would mean that, after an initial reference measurement, BIS is a valuable method to assess the BAC. Since this technology is non-invasive and can be measured using conductive electrodes, capacitive electrodes or inductive measurement systems, this allows to not only establish devices for traffic control, but also for unobtrusive sensors in cars. Although bioimpedance cannot totally compete with the existing methods, this technique can be used as a complementary measure.

Additional trials with more participants (of both sexes, and of varying weight) are required. To avoid possible bias, a change in role with respect to water and/or alcohol consumption, and a repeat measurement for both groups, should be taken into account. In case of a contactless device, analysis focusing on the signal-to-nose ration (SNR) is necessary because it is known that, for example, inductive bioimpedance measurements can achieve only low SNRs (between 20 dB and 50 dB) [9].

ACKNOWLEDGMENT

Michael Czaplik was funded by the Medical Faculty at RWTH Aachen University.

Special thanks go to Angela Sudhoff and Verena Deserno from the Clinical Trials Center in Aachen (CTC-A) and the Center for Laboratory Diagnostics at the University Hospital Aachen.

REFERENCES

[1] The Federal Statistical Office of Germany, *Unfallentwicklung auf deutschen Straßen 2010*, T.F.S.O. of Germany, Ed., 2011.

[2] T.R. Leffingwell, N.J. Cooney, J.G. Murphy, *et al.*, "Continuous Objective Monitoring of Alcohol Use: Twenty-First Century Measurement Using Transdermal Sensors," *Alcoholism: Clinical and Experimental Research*, vol. 37, pp. 16–22, 2013.

[3] Y. Omura and K. Kojima, "Spectroscopic Study and Analysis of the Impact of Alcohol Intake on Bio-Impedance of the Human Body," *IEEE Sensors*, pp. 1648–1651, 2011.

[4] S.B. Karch, *Forensic Issues in Alcohol Testing*. CRC Press, 2010.

[5] U. Schröder, "Vergleich der Blutalkoholkonzentration mit der Atemalkoholkonzentration nach mässigem Alkoholkonsum," PhD thesis, RWTH Aachen University, 2004.

[6] S. Grimnes, *Bioimpedance and Bioelectricity Basics*. Academic Press, 2000.

[7] G. Medrano, F. Eitner, M. Walter, *et al.*, "Model-Based Correction of the Influence of Body Position on Continous Segmental and Hand-to-Foot Bioimpedance Measurements," *Medical & Biological Engineering & Computing*, vol. 48, pp. 531–541, 2010.

[8] L. Beckmann, C. Neuhaus, G. Medrano, *et al.*, "Characterization of textile electrodes and conductors using standardized measurement setups," *Physiological Measurement*, vol. 31, pp. 233–274, 2010.

[9] D. Teichmann, J. Foussier, J. Jia, *et al.*, "Noncontact Monitoring of Cardiorespiratory Activity by Electromagnetic Coupling," *IEEE Transactions on Biomedical Engineering*, vol. 60, pp. 2142–2152, 2013.

Batteries

Lecture Notes on Impedance Spectroscopy, Volume 5 – Kanoun (Ed.)
© *2015 Taylor & Francis Group, London, ISBN 978-1-138-02754-1*

Thermal characterization of Li-ion batteries: Accelerating Thermal Impedance Spectroscopy

Peter Keil, Katharina Rumpf & Andreas Jossen
Institute for Electrical Energy Storage Technology, Technische Universität München, Munich, Germany

ABSTRACT: Thermal Impedance Spectroscopy (TIS) is a cost-effective and reliable method for identifying thermal battery parameters. However, the measurements are time-consuming, requiring approximately one day for a single battery. In this paper, several methods for accelerating TIS measurements are investigated. Reducing the number of frequencies, the number of oscillations per frequency, and a separate identification of heat exchanged with environment shortens TIS measurements markedly. With an optimized TIS procedure that is based on an evaluation of not only alternating signals but also direct components, measurement time can be reduced from one day to less than five hours. A comparison of results from conventional and optimized TIS procedure shows that this can be achieved without a considerable loss in accuracy.

Keywords: Thermal Impedance Spectroscopy; parameter identification; Li-ion batteries; time domain analysis

1 INTRODUCTION

Measuring thermal parameters of Li-ion cells is crucial for optimizing the thermal design of battery systems with respect to lifetime and safety issues. The thermal parameters of interest are heat capacity, thermal conductivity, and heat exchange between the cell's surface and the environment due to radiation and convection. Traditionally, heat capacity is obtained by calorimeter measurements and thermal conductivity is obtained by heat flux or Xenon-Flash measurements [1]. Disadvantages of these methods are the requirement of expensive measurement devices and the destruction of the cell for thermal conductivity measurements.

Barsoukov et al. [2] introduced Thermal Impedance Spectroscopy (TIS) as a nondestructive method for examining thermal parameters of entire battery cells. They apply a sinusoidal heat excitation to cylindrical cells with a heating band wound around the cell's housing. By measuring the cell's responding surface temperature and transferring both excitation and response signal to the frequency domain, the thermal impedance of the cell is calculated for different excitation frequencies. A plot of the thermal impedance for the measured frequencies in the complex plane yields the thermal impedance spectrum of the cell. The thermal parameters are determined by fitting a thermal model to the characteristic impedance spectrum. Schmidt et al. [3] and Fleckenstein et al. [4] have enhanced the TIS method by using internal irreversible losses of the battery instead of an external heating band to induce a sinusoidal heat generation.

As heat conduction is rather slow, only frequencies in the millihertz range provide meaningful information. The frequencies applied in the three above-mentioned references are all in an interval between 100 mHz and 0.1 mHz. As each excitation frequency has to be applied for several oscillations to achieve steady conditions, a TIS measurement lasts approximately one day [3], [4]. Since one day per measurement is too long for characterizing and comparing a larger number of battery cells, different methods for accelerating TIS measurements are presented and evaluated in this paper.

2 BASIC PRINCIPLES OF TIS MEASUREMENT

Before describing the approaches for acceleration, the fundamentals of conventional TIS measurements are explained.

2.1 Thermal impedance definition

To describe the heat transfer behavior of a battery for a certain excitation frequency f, a thermal impedance Z_{th} is used. This impedance, representing the temperature response $T(\omega t)$ to a sinusoidal heat excitation $Q(\omega t)$, calculates in the frequency domain as:

$$Z_{th}(\omega t) = \frac{T(\omega t)}{Q(\omega t)} = \frac{\hat{T} \cdot e^{j(\omega t + \phi_T)}}{\hat{Q} \cdot e^{j(\omega t + \phi_Q)}} = \frac{\hat{T}}{\hat{Q}} \cdot e^{j(\phi_T - \phi_Q)}, \quad \omega = 2\pi f \tag{1}$$

As frequencies in the millihertz and sub-millihertz range are examined, the evaluation of measurement data in the time domain instead of the frequency domain is beneficial. This is also discussed later in chapter III.A. From time domain data, impedances Z_{th} are derived by a calculation of amplitude ratio ($\hat{T} / \hat{Q}$) and phase delay ($\Delta\phi = \phi_T - \phi_Q$).

Thermal impedances for different frequencies result in a characteristic impedance spectrum as illustrated in Figure 1.

2.2 Heat generation

To obtain Z_{th}, a sinusoidal heat flow $Q(t)$ has to be applied to the battery. Preferably, this is performed by using irreversible losses of the battery itself that provoke internal heat generation. Applying a sinusoidal current to the battery would cause a charge and a discharge half cycle, which both last up to several minutes and hours for low frequencies in the millihertz and sub-millihertz range. Consequently, the battery's state of charge changes considerably.

This can be avoided by using an alternating charge and discharge carrier that is modulated with the test frequency. Figure 2 illustrates the method in which current profiles for our TIS measurements are composed. A low-frequency sinusoid provides the required test frequency in the range of millihertz or microhertz. This signal is multiplied with an alternating, rectangular profile to switch between charge and discharge periods every 3 s. Thus, the battery's state of charge remains constant while, at the same time, a sinusoidal heat generation can be achieved.

An offset in the low frequency sinusoid signal leads to a permanent generation of heat losses to prevent the battery from cooling down to environment temperature. The excitation of the battery is performed with a BaSyTec battery test system. Figure 5a and Figure 5b show current and voltage measurements of a TIS experiment.

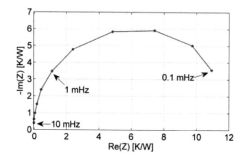

Figure 1.   Typical TIS spectrum of a cylindrical 18650 Li-ion battery.

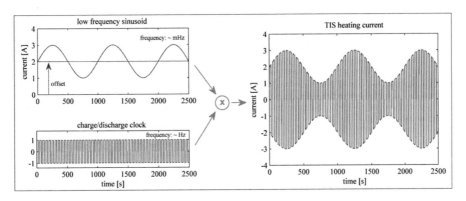

Figure 2.   Illustration of heating current composition for TIS experiments.

Figure 3.   PerkinElmer A2TMPI 334 thermopile sensor.

2.3   *Heat loss calculation*

Electrical measurement values, such as current, voltage, charge balance (Ah), and energy balance (Wh), are recorded by the battery test system every 500 ms. Charge and energy balance are updated inside the test system every few milliseconds. From the recorded data, heat losses are calculated after the measurement. For each pair of charge and discharge pulses, the energy loss is calculated. An interpolation between measurement points assures that the evaluated end point of a pulse pair has the same charge content as the start point. In this case, the energy loss equals the difference in Wh balance between start and end point. Dividing these energy losses by the duration of the pulse pair yields the required heating power. As the duration of the pulse pairs is short compared to an entire period of the low-frequency sinusoid, the heat calculation provides a smooth and accurate heat generation curve which can be seen later in Figure 5c.

2.4   *Temperature measurement*

In addition to the heating power curve, a surface temperature curve is required as the second input for thermal impedance calculation. To overcome drawbacks of contact-based temperature sensors, radiation-based temperature sensors are used for surface temperature measurement, as described in [5]. The PerkinElmer A2TMPI 334 sensor used in this study is shown in Figure 3.

The sensor has an integrated data processing unit and covers a temperature range from −20 °C to 100 °C. It converts thermopile voltages of a few microvolts into an output voltage between 0 V and 5 V. The output voltage is digitalized by a 24 bit analog-to-digital converter and the result is transmitted via an USB interface. Temperature data are recorded synchronously by the battery test system. Figure 5d shows raw and median-filtered surface temperature values.

2.5   Impedance calculation

As formulated in Equation (1), thermal impedance Z_{th} has to be calculated from surface temperature and heat losses. Firstly, measurement data are subdivided into sections for each frequency. Then, surface temperature curve and heat losses are approximated by an analytical expression that has been presented in [5]:

$$y(t) = O + A_1 \cdot sin(2\pi ft + \alpha_1) + A_2 \cdot sin(4\pi ft + \alpha_2) \qquad (2)$$

with test frequency f, offset O, amplitudes A_1, A_2, and phase angles α_1, α_2. This expression leads to an excellent approximation of measurement curves (see Figure 4).

The ratio of A_1 values and the difference of α values from heating and temperature curves represent amplitude and phase of the thermal impedance Z_{th} for each frequency f.

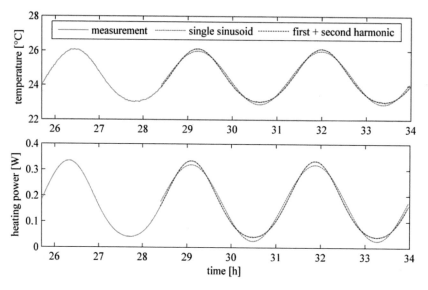

Figure 4.   Approximation of the last two thirds of experimental data with a single frequency sinusoid and with a combination of two sinusoids (first + second harmonic).

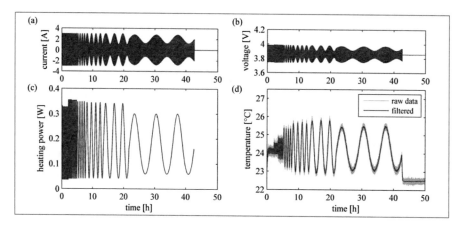

Figure 5.   Measurement data of a TIS experiment comprising (a) current, (b) terminal voltage, (c) computed heat losses, and (d) surface temperature (raw and median-filtered thermopile data).

$$Z_{th,1}(j\omega) = \frac{A_{1,T}}{A_{1,P}} \cdot e^{j(\alpha_{1,T} - \alpha_{1,P})} \qquad (3)$$

This impedance calculation can also be performed for the second harmonics ($2f$) by evaluating the corresponding pairs of A_2 and α_2 values.

The impedance spectrum obtained from fundamental frequencies f serves as the basis for thermal parameter identification; the spectrum obtained from harmonics is used for consistency checks.

2.6 Measurement setup

Figure 6 shows a schematic representation of the TIS measurement setup. The cylindrical Li-ion cell rests on two stands, which are designed in such a way that the bearing area is minimized to avoid distortions by thermal conduction. The thermopile sensor in located below to the cell in a distance of less than 10 mm. The battery is clamped between two pairs of spring contacts. This provides a four wire connection to the battery test system, which is required for precise voltage measurements without any voltage drops along the connecting leads.

Due to the small contact area between the contact tips and the caps of the cell, the heat flow from the cell to the connecting leads is minimized. A thermal camera is used to verify the homogeneous surface temperature of the cell. Deviations were less than 0.1 K and no temperature drop at the caps were observed. Moreover, the heat produced by the contact resistances of the small contact tips did not increase the temperature of the caps, as this heat was removed by the connecting leads.

2.7 Thermal battery model

In order to extract thermal battery parameters from a measured impedance spectrum, a thermal model of the battery is necessary. This model reproduces the effects of heat capacity, thermal conductivity, heat exchange with environment, and internal losses generation. Transient simulation is then used to calculate an impedance spectrum for the battery model which is subsequently employed for parameter identification.

Since all our TIS experiments are performed with cylindrical 18650 batteries inside a temperature chamber, homogeneous environment temperatures can be assumed. As generated heat losses inside the cells are rather low (<0.5 W) and the temperature increase during TIS experiments accounts only for a few degrees centigrade, the battery exhibits a homogeneous surface temperature. This was verified by thermal images of an infrared camera. The uniform temperature distribution allows a reduction of the battery model to a 1D heat transfer problem in radial direction [5]. Consequently, this leads to a neglect of heat exchange at the caps. This is acceptable, as the heat conduction through the contact tips is very low due to the miniscule contact area. Moreover, the amount of heat transferred at the caps due to convection and radiation will automatically be part of to the heat exchanged at the cylindrical

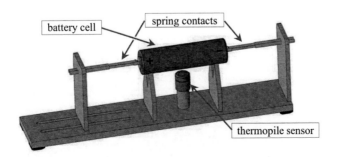

Figure 6.   Schematic representation of TIS measurement setup.

43

surface, when solving the power balance equations between battery and environment shown in chapter II.G.5.

In order to simulate the radial heat transfer numerically with a finite differences method, the battery has to be discretized in radial direction. Figure 7 shows a cylindrical battery, which is discretized equidistantly into hollow cylinders. Moreover, it illustrates the reduction to one relevant dimension with heat transfer only in radial direction.

Each discretization element i consists of an element temperature T_i and a lumped heat capacity C_{p_i}. Neighboring elements are linked by a thermal resistance $R_{th\_i,i+1}$, representing thermal conductivity. Heat exchange with environment comprises radiation Q_{rad} and convection Q_{conv}. Internal heat generation is expressed by irreversible losses P_{v_i}. For our simulations, a discretization with $N = 10$ is used, which has turn out to be a good compromise between accuracy and computational effort.

a. *Heat capacity*
The heat capacity of the entire battery C_p equals the product of mass m and specific heat capacity c_p of the battery:

$$C_p = m \cdot c_p \tag{4}$$

The lumped heat capacity C_{p_i} of each discretization element with volume V_i is proportional to its ratio of the entire battery volume V ($\pi r^2 h$), where r is the radius and h the height of the battery. For a uniform discretization, the thickness of all hollow cylinders is identical (r/N). This leads to the following volume and heat capacity shares for the discretization elements:

$$V_i = \left(i \cdot \frac{r}{N} \right)^2 \pi h - \left((i-1) \cdot \frac{r}{N} \right)^2 \pi h = \frac{2i-1}{N^2} r^2 \pi h \tag{5}$$

$$C_{p\_i} = \frac{V_i}{V} \cdot C_p = \frac{2i-1}{N^2} \cdot C_p \tag{6}$$

b. *Thermal conduction*
Exchanged heat $Q_{i,i+1}$ between two neighboring discretization elements with lumped heat capacities C_{p_i} and C_{p_i+1} can be expressed by Fourier's law as:

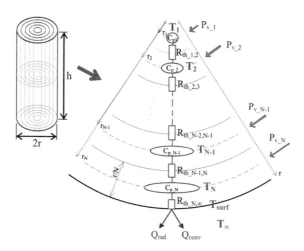

Figure 7. Discretization of a cylindrical battery for finite differences simulation and reduction to a radial 1D heat transfer problem.

$$Q_{i,i+1} = R_{th\_i,i+1} \cdot \left(T_i - T_{i+1}\right) \tag{7}$$

The thermal resistance $R_{th\_i,i+1}$ for the heat flow through the cylinder segment between two neighboring heat capacities is calculated as in [6]:

$$R_{th\_i,i+1} = \frac{ln(r_{i+1}/r_i)}{\lambda \cdot 2\pi h} \quad \forall i \in [1, N-1] \tag{8}$$

where r_i and r_{i+1} are the inner and outer radii of the heat transfer distance between the two lumped heat capacities. λ is the thermal conductivity of the battery in radial direction.

c. *Heat exchange with environment*

As the battery exchanges heat with the environment, radiative and convective heat flows are applied to the outermost discretization element. These two boundary conditions depend on the battery's surface temperature T_{surf} and the environment temperature T_∞. T_{surf} is computed based on the thermal resistance $R_{th\_N,\infty}$, which represents the heat conduction through the outer half of the outermost discretization element and is calculated similarly to the other resistances $R_{th\_i,i+1}$ as explained above.

Radiated heat from the battery follows the Stefan-Boltzmann law and is obtained as:

$$Q_{rad} = A_{surf} \cdot \varepsilon \cdot \sigma \cdot F \cdot \left(T_{surf}^4 - T_\infty^4\right) \tag{9}$$

where A_{surf} is the battery's surface area ($2\pi rh$), ε is the emissivity, σ is the Stefan-Boltzmann constant, and F is the view factor between battery and environment. Since the battery is a convex object in a larger enclosure, F can be set to 1 [7]. As the battery's surface is covered with black insulation tape for the TIS measurements, ε is close to 1. Here, it is estimated to be 0.95 as in [8].

For values of T_∞ close to room temperature, Q_{rad} can be simplified by the following expression [9]:

$$Q_{rad} = \alpha_{rad} \cdot A_{surf} \cdot \left(T_{surf} - T_\infty\right) \tag{10}$$

where α_{rad} is a constant heat transfer coefficient.

In addition to radiation, heat is also exchanged with environment by natural or forced convection. This additionally emitted heat computes as:

$$Q_{conv} = \alpha_{conv} \cdot A_{surf} \cdot \left(T_{surf} - T_\infty\right) \tag{11}$$

where α_{conv} is the connective heat transfer coefficient, which is assumed to be independent of temperature.

Adding up Equations (10) and (11) leads to a single heat exchange coefficient α_{env} that combines radiation and convection and yields the total amount of heat exchanged with environment:

$$Q_{env} = \alpha_{env} \cdot A_{surf} \cdot \left(T_{surf} - T_\infty\right) \tag{12}$$

d. *Irreversible heat losses*

The implementation of irreversible losses P_v, used for heating the battery, completes the simulation model. As only inconsiderable temperature gradients occur inside the battery during the applied low-frequency heating currents, losses are assumed to be uniformly distributed over the volume of the cell. Thus, P_{v_i} is proportional to V_i. In analogy to Equation (6), P_{v_i} calculates as:

$$P_{v\_i} = \frac{v_i}{v} \cdot P_v = \frac{2i-1}{N^2} \cdot P_v \tag{13}$$

e. *Transient thermal simulation*
To simulate transient thermal behavior of the battery, a heat balance is formulated for each discretization element:

$$C_{p_i} \cdot \dot{T}_i = Q_{i-1,i} - Q_{i,i+1} + P_{v_i} \quad \forall i \in [1, N-1] \tag{14}$$

$$C_{p_N} \cdot \dot{T}_N = Q_{N-1,N} + P_{V_N} - Q_{rad} - Q_{conv} \tag{15}$$

Composing these heat balances for all discretization elements to a system of equations yields:

$$\mathbf{C_p} \cdot \dot{T} = \mathbf{W_{cond}} \cdot T - Q_{env} + P_v \tag{16}$$

where T is the vector of all element temperatures, $\mathbf{C_p}$ is the diagonal heat capacity matrix, $\mathbf{W_{cond}}$ is the heat conduction matrix, vector Q_{env} (with just one entry at Nth position) represents the sum of radiative and convective heat exchange with the environment, and vector P_v contains the generated heat. This equation is solved numerically over time and delivers the temperature response of the battery to a specified heating signal.

f. *Parameter identification*
The implemented battery model is the basis for the identification of thermal cell parameters. With the battery model, TIS measurements are simulated. The same sinusoidal heat excitation as in the experiment is applied to the thermal battery model. Simulation results deliver thermal impedances for each frequency, which form an entire impedance spectrum. The ability to simulate TIS measurements allows rapid creation of impedance spectra for arbitrary thermal cell parameters.

A least-squares optimization routine is employed to systematically adapt the heat capacity, thermal conductivity, and convective heat exchange coefficient of the simulation model until a good agreement between measured and simulated impedance spectra is achieved. Final values of the parameter variation process represent the thermal parameters of the real battery. Figure 8 compares an impedance spectrum from measurement data with an impedance spectrum derived from the result values of the optimization process. As a good agreement

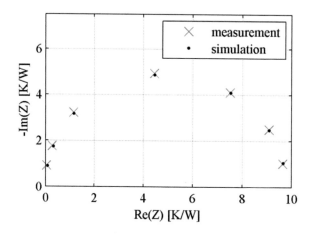

Figure 8.   Comparison of thermal impedance spectra from measurement and simulation.

46

between both spectra can be achieved, the utilized battery model is appropriate for thermal parameter identification.

3   METHODS FOR ACCELERATING TIS MEASUREMENTS

The conventional TIS procedure leads to a measurement time of about one day per battery. Since one day per measurement is too long for characterizing a larger number of battery cells, in this paper, we present four methods for accelerating TIS measurements:

A. Reduction of oscillations per frequency
B. Reduced set of frequencies
C. External value for heat exchange with environment
d. Evaluation of alternating and direct signal components

In the following paragraphs, the main ideas of the four methods are presented. An evaluation of the acceleration techniques is provided in the results and discussion sections.

A.   *Reduction of oscillations per frequency*
The lower the test frequency becomes, the more time is required for the respective section of a TIS measurement run. For the time-consuming measurements of frequencies below 1 mHz, three or four oscillations are performed in [3], of which two oscillations are evaluated afterwards for impedance calculation.

Due to the low frequencies of TIS measurement, data processing with curve fitting in the time domain has replaced methods in the frequency domain, such as FFT. This allows an evaluation of a non-integer number of oscillations without the need of applying windowing functions. Consequently, less measurement data have to be discarded at the beginning of each new test frequency, when there occurs a certain transition period, until new steady conditions are reached.

To accelerate TIS measurements, a reduction of oscillations per frequency to only two oscillations for frequencies below 1 mHz can decrease measurement duration by approximately one third.

Measurements with 18650 Li-ion cells showed that the transition period after changing the test frequency did not last for more than 15 minutes. Discarding the first 20 minutes of each test frequency, 1.25–1.88 oscillations remain for further evaluation. Figure 9 shows the excellent agreement between measurements and approximated curves for the data section used for impedance calculation.

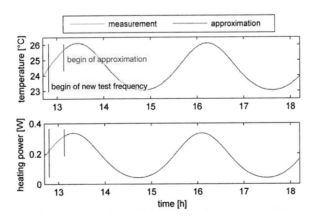

Figure 9.   Approximation of temperature and heating power curve for 0.1 mHz when two oscillations are performed and the first 20 minutes are discarded.

B. *Reduced set of frequencies*

Since not all excitation frequencies provide the same benefit for thermal parameter identification, certain frequencies can be eliminated without altering noticeably the thermal cell parameters identified. As the lowest test frequencies are the most time-consuming ones, they are the preferred candidates for elimination.

Figure 10 shows how the available impedance spectrum shrinks, when up to five frequencies are eliminated and the spectrum reaches only down to 1 mHz. It has to be determined down to which frequency the measured spectrum has to reach in order to maintain high precision. With this information in mind, a reduction of measured frequencies per decade might also help to reduce measurement time but is not examined further in this paper.

C. *External value for heat exchange with environment*

First measurements have shown that a precise identification of the coefficient for heat exchange with environment is crucial. This coefficient correlates with the diameter of the semicircle of the impedance spectrum. For a precise reproduction of this diameter, TIS measurement data in the low frequency range (<1 mHz) are required. Predicting the diameter from impedance values of solely higher frequencies leads to considerable deviations.

However, a precise reproduction of the impedance spectrum without low frequency data is possible, when the coefficient for heat exchange with environment is determined by a separate measurement. This can be achieved by heating up the cell with an alternating, rectangular charge/discharge profile, which is illustrated in Figure 11.

After having reached steady-state conditions ($\dot{T} = 0$), Equation (16) leads to the result that there must be equilibrium between internally generated heat and heat exchanged with the environment by radiation and convection:

$$Q_{env} = P_v \tag{17}$$

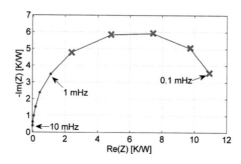

Figure 10.   Elimination of low frequencies that lead to a reduced impedance spectrum for thermal parameter identification.

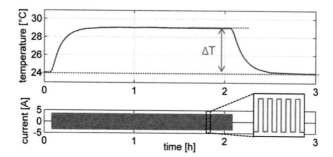

Figure 11.   Battery heating with alternating charge/discharge profile until steady-state conditions are reached.

48

From this equilibrium, the heat exchange coefficient α_{env} can be calculated after transformation of Equation (12):

$$\alpha_{env} = \frac{P_v}{A_{surf} \cdot \left(T_{surf} - T_\infty\right)} \tag{18}$$

The heat exchange coefficient is then removed from the parameter identification process, so that only heat capacity and thermal conductivity remain for identification.

By providing a predefined value for the heat exchange coefficient, fewer low frequencies are required for precise parameter identification and TIS measurements can be shortened. However, the separate measurement of the heat exchange coefficient requires additional measurement time of about three hours for an 18650 Li-ion cell. Consequently, this acceleration approach is particularly advantageous, when several cells of the same type are investigated, as the heat exchange coefficient has to be determined only once.

D. *Evaluation of alternating and direct signal components*

In conventional TIS data analyses, only alternating signal components are evaluated. Taking also direct signal components into evaluation enables a calculation of the heat transfer coefficient for heat exchange with environment without additional measurements.

As no negative irreversible heat losses can be generated inside the battery, the sinusoidal heat excitation of TIS measurements always contains an offset. This offset ΔP represents the direct component of the heat excitation signal (see Figure 12).

As a consequence, this leads to a cell temperature which always lies above ambient temperature. In steady-state conditions, the average battery temperature represents the direct component of the sinusoidal temperature curve (right subplot of Figure 12).

To calculate the coefficient for heat exchange with environment, the difference ΔT between average cell temperature and environment temperature and the average heating power ΔP are part of the following equation:

$$\alpha_{env} = \frac{P_v}{A_{surf} \cdot \Delta T} \tag{19}$$

Using direct signal components to separately determine the heat exchange with environment saves time for additional measurements. It also provides a precise value for the heat exchange coefficient and, thus, reduces the number of low frequencies required for the parameter identification process. A minimum frequency of 1 mHz instead of 0.1 mHz reduces measurement time to less than 5 hours.

Consequently, the evaluation of alternating as well as direct signal components provides a maximum time reduction for TIS measurements.

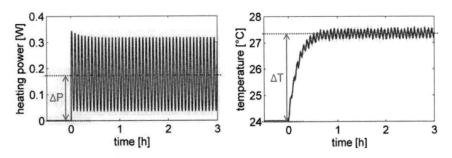

Figure 12. Direct components under steady-state conditions used for the calculation of the coefficient for heat exchange with environment.

4 RESULTS

Based on thermal parameters identified by a reference TIS measurement, the effectiveness of the four illustrated acceleration techniques is presented in Table 1. For the reference measurement, different sets of measurement frequencies were picked and processed to evaluate the robustness of the data processing routines. The parameters heat capacity and heat transfer coefficient for heat exchange with the environment can be identified with low uncertainty: The estimated standard deviations are less than 1% of the mean value. In contrast, thermal conductivity exhibits a larger standard deviation of 16% indicating higher uncertainties. These uncertainties are important for the following evaluation of acceleration methods for TIS measurements.

By a reduction of oscillations per frequency in the low-frequency domain from 3 to 2, method A shortens the time span for one TIS measurement by 25% to 18.4 h. Heat capacity and heat exchange with environment still have similar values. For heat capacity, the standard deviation from several evaluation runs increases slightly but stays below 1%. Regarding thermal conductivity, the minor changes in mean value and standard deviation of 0.1 W/mK lie clearly below the uncertainties of the reference measurement expressed by standard deviation. Consequently, thermal parameters confirm that the reduced number of oscillations has practically no influence on accuracy.

Method B shows the effects of a reduced range of frequencies. With an increasing value for the minimum frequency, thermal parameters diverge increasingly from the reference values. However, the first two cases listed in Table 4 demonstrate that not all low frequencies are necessary for accurate parameter identification. Choosing 0.25 mHz instead of 0.1 mHz as the minimum frequency provides a time saving of approximately 9 h (50%) compared to method A and 16 h (60%) compared to the reference scenario. This is possible without relevant loss in precision. A further omitting of low frequencies leads to increased deviations of up to 1.4% in heat capacity, 6% in heat exchange with environment, and 80% in thermal conductivity. Thus, choosing a minimum frequency of 0.4 mHz or higher for conventional TIS measurements prevents any reliable parameter identification.

Method C represents an optimized TIS procedure. The separate identification of heat exchange with environment results in accurate heat capacity and thermal conductivity identification without the need of frequencies below 1 mHz. All thermal parameters lie

Table 1. Quality of thermal parameter identification of reference measurement and accelerated measurements.

	Measurement setup			Duration	Thermal parameter identification		
	Maximum frequency (mHz)	Minimum frequency (mHz)	Osciliations per frequency (f < 1mHz)	Total time (h)	Heat capacity (J/K)	Thermal conductivity (W/mK)	Heat exchange with environment (W/m²K)
reference	10.0	0.10	3	25	43.6 ± 0.2	3.2 ± 0.5	27.8 ± 0.1*
method A	10.0	0.10	2	18.4	43.6 ± 0.4	3.1 ± 0.6	27.8 ± 0.1*
method B	10.0	0.16	2	12.8	43.6	3.0	27.8
	10.0	0.25	2	9.3	43.6	3.1	27.7
	10.0	0.40	2	7.1	43.8	4.3	27.3
	10.0	0.63	2	5.7	43.9	5.2	26.7
	10.0	1.0	2	4.8	44.2	5.7	26.0
method C	10.0	0.10	2	3 + 4.8	43.7	3.3	27.8
method D	10.0	0.10	2	4.8	43.7	3.5	27.8

*Mean ± standard deviation comprising several evaluations to illustrate uncertainties due to the parameter identification routines.

close to the reference values and deviations are all within the standard deviation. Method C leads to a further time reduction of 13.6 h (74%) for the TIS measurement compared to the measurement according to method A. However, additional 3 h are required for determining the heat exchange coefficient for convection and radiation at steady-state conditions. This extra measurement is necessary for every new measurement setup and new type of cell. When investigating only one cell, almost 8 h are necessary for the entire measurement procedure. This is approximately one third of the duration of the reference measurement.

Finally, method D demonstrates that TIS data contain more meaningful information for thermal parameter identification than utilized by conventional TIS techniques. Evaluating direct signal components facilitates a precise identification of the heat exchange with environment. This leads to the same reduction of TIS measurement time as method C. However, method D achieves this without any additional measurements. It enables the parameter identification for a single cell within less than five hours. The last row of Table 1 shows that the values of heat capacity and heat exchange with environment are similar to the reference measurement. The difference in thermal conductivity of 0.3 W/mK is still lower than the standard deviation of 0.5 W/mK of the reference measurement. In total, method D reduces measurement time by about 80% while differences in parameter values remain negligible.

5 CONCLUSION & PRACTICAL APPLICATION

TIS has already been presented in [3], [4] as a cost-effective and reliable method for the identification of thermal cell parameters. Moreover, we could show that there is a large potential of accelerating TIS measurements.

In addition to an optimization of the set of examined frequencies and the number of oscillations, beneficial effects of a separate determination of the heat exchange with environment were identified. Either by an additional heating measurement or by an evaluation of direct signal components, a precise identification of the heat exchange coefficient is possible. This reduces markedly the required range of frequencies for the accurate identification of heat capacity and thermal conductivity.

Finally, we achieved a reduction of measurement time from one day to less than five hours without a considerable loss in accuracy. This makes TIS convenient for frequent use even for larger amounts of batteries. An identification and comparison of thermal parameters might become a standard routine for the assessment of new cells before assembling them to battery systems.

The improved TIS measurement procedure could also be adapted and applied to entire battery packs. Heat exchange with the cooling fluid could be analyzed this way in order to optimize cooling architecture and operational strategies.

REFERENCES

[1] H. Maleki, S. Al Hallaj, J.R. Selman, *et al.*, "Thermal Properties of Lithium-Ion Battery and Components," *Journal of the Electrochemical Society*, vol. 146, no. 3, pp. 947–954, 1999.
[2] E. Barsoukov, J.H. Jang, and H. Lee, "Thermal impedance spectroscopy for Li-ion batteries using heat-pulse response analysis," *Journal of power sources*, vol. 109, no. 2, pp. 313–320, 2002.
[3] J.P. Schmidt, D. Manka, D. Klotz, *et al.*, "Investigation of the thermal properties of a Li-ion pouch-cell by electrothermal impedance spectroscopy," *Journal of Power Sources*, vol. 196, no. 19, pp. 8140–8146, 2011.
[4] M. Fleckenstein, S. Fischer, O. Bohlen, *et al.*, "Thermal Impedance Spectroscopy—A method for the thermal characterization of high power battery cells," *Journal of Power Sources*, vol. 223, pp. 259–267, 2013.

[5] P. Keil, K. Rumpf, and A. Jossen, "Thermal Impedance Spectroscopy for Li-Ion Batteries with an IR Temperature Sensor System," in *The 27th World Battery, Hybrid and Fuel Cell Electric Vehicle Symposium and Exhibition*, 11/2013.

[6] F. Kreith, R. Manglik, and M. Bohn, *Principles of Heat Transfer*, Cengage Learning, Ed. Stanford, 2011, vol. 7.

[7] W. Polifke and J. Kopitz, *Wärmeübertragung*, Pearson Studium, Ed. Munich, 2005, vol. 1, p. 115.

[8] T. Polak and C. Pande, *Engineering measurements: methods and intrinsic errors*, Professional Engineering, Ed. Michigan, 1999, vol. 1, p. 67.

[9] A. Jossen and W. Weydanz, *Moderne Akkumulatoren richtig einsetzen*, Ubooks, Ed. Neusäß, 2006, vol. 1, p. 22–23.

Lecture Notes on Impedance Spectroscopy, Volume 5 – Kanoun (Ed.)
© *2015 Taylor & Francis Group, London, ISBN 978-1-138-02754-1*

Physico-chemical modelling of Li-ion batteries: Parameter analysis in the frequency domain

Simon Erhard, Franz Spingler, Alexander Rheinfeld, Stephan Kosch, Katharina Rumpf & Andreas Jossen
Institute for Electrical Energy Storage Technology (EES), Technische Universität München, Munich, Germany

ABSTRACT: In this article an impedance based analysis of a physico-chemical model is presented. This method facilitates identifying the influence of various internal cell parameters on the resulting impedance spectrum. Thus, a precise correlation between a particular parameter or even a set of parameters and the observed frequency characteristics of a full cell can be revealed. For the specific case investigated here, the anode side can be identified as having the dominating impact on the impedance characteristics of the cell. Moreover, the effect of varying diffusion coefficients is astonishingly negligible compared to the particle radius or reaction rates at each electrode side.

Keywords: Li-ion batteries; physico-chemical modelling; parameter analysis; frequency domain

1 INTRODUCTION

Publications on the modelling of electrochemical impedance spectroscopy can be divided into two major approaches: First, the analogous representation of system behaviour via Electrical Equivalent Circuits (EEC) and second, the modelling of system behaviour by a set of physico-chemical equations. There is broad agreement among many authors that impedance modelling based on EEC only allows for limited phenomenological insight [1]–[4]. This constraint emphasizes the need for a more fundamental approach to describe impedance behaviour.

The governing differential equations of physico-chemical models are generally time dependent. Impedance, however, is defined in the frequency domain. Among previously published research work, the governing model equations were mostly transformed into frequency domain and were then used to directly calculate the impedance from analytical expressions. Other research groups have calculated the stimulation response in time domain and then used techniques like Fast Fourier Transform (FFT) to obtain impedance data from resulting harmonic oscillations [5]–[7]. An advantage of calculation in the time domain is the possibility of studying non-periodic transients and nonlinear behaviour [5].

In this work, we perform a sensitivity analysis of selected parameters of a commercial 26650 LiFePO$_4$/graphite cell and investigate their effect on the simulated impedance spectrum. Basic values such as layer thickness and particle radii are taken from literature and preceding measurements. The model implemented within the commercial Finite Element Method (FEM) software COMSOL Multiphysics is then solved in the frequency domain. To demonstrate the capabilities of this method, variations in state of charge, particle radius, solid state diffusion coefficient and reaction rate are analysed. These parameters evoke characteristic and also unusual properties of the observed impedance spectrum.

2 ELECTROCHEMICAL MODEL

Based on Newman's well-known modelling approach [8]–[10] the impedance of a commercial cell is described. This approach combines concentrated solution theory, porous electrode theory and Butler-Volmer kinetics to form a set of coupled partial differential equations.

2.1 *Geometry*

The model geometry is described in two dimensions. In x-direction the flux of ions and electrons accounts for charge transport perpendicular to the cell layers. The influence of ohmic drops within the current collectors is assumed to be zero, thus only three domains are taken into account. These are the anode (graphite), a porous separator and the cathode (LiFePO$_4$). The solid diffusion within the electrode's active material particles is calculated in an additional pseudo-dimension in spherical coordinates. So at every point x within an electrode domain a second dimension r is used to describe this flux directed to or away from the particle's centre. The dimensions are coupled at the particle's surface. A binary electrolyte (one salt in one solvent) is assumed, whereas only the cation flux is described, since the anions do not contribute to the electrochemical reaction of the cell. The subscript + indicates the aforementioned cationic species (Li$^+$).

2.2 *Mass balance*

Within the liquid phase a mass balance for the Li$^+$ is given, which takes flux due to diffusion and migration into account. These equations assume a binary electrolyte consisting of one salt in one solvent. In most Li-ion batteries more than one solvent is used. However, it is a practical assumption for small stimulation signals to take only one solvent species into consideration [10]. The concentration of Li$^+$ within the electrolyte (subscript l) in mol/m^3 in x-direction is given by the mass balance

$$\varepsilon_l \frac{\partial c_l}{\partial t} = -\nabla \mathbf{N}_+ + R \tag{1}$$

and the flux of Li$^+$

$$\mathbf{N}_+ = -D_l \nabla c_l + \frac{\mathbf{i}_l t_+}{F} \tag{2}$$

Thus, the change in concentration due to ion flux is defined by the diffusion coefficient D_l of the salt, the current within the liquid phase $\mathbf{i}_l(A/m^2)$ and the transference number t_+. The reaction term describes charge transfer reactions at the electrode/electrolyte interface. Therefore, this term is zero in the separator domain, since there is no charge transfer reaction. The volume fraction of the electrolyte ε_l can be considered as the domain's porosity.

Since the observed length scales within that model are above 0.1 μm, the assumption of electroneutrality is considered [10].

2.3 *Charge balance*

In the solid phase (subscript $_s$) the flux of electrons is described by Ohm's law

$$\mathbf{i}_s = -\sigma \nabla \phi_s \tag{3}$$

whereas ϕ_s describes the potential of the electrode with respect to a certain reference (V vs. Li/Li$^+$) and represents the electronic conductivity of the electrode in S/m. Since the separator is an insulating material, this equation is only valid for the electrode domains. In order to assure charge conservation within both phases, the following charge balance is assumed for a porous electrode domain

$$\nabla \mathbf{i}_s = -\nabla \mathbf{i}_l \qquad (4)$$

whereas

$$\nabla \mathbf{i}_l = 0 \qquad (5)$$

accounts for the separator domain. Within the liquid phase the current-potential relation is given by a modified Ohm's law

$$\mathbf{i}_l = -\kappa \nabla \phi_l + \frac{2\kappa RT}{F}\left(1 + \frac{d \ln f_\pm}{d \ln c_l}\right)(1 - t_+)\nabla \ln c_l \qquad (6)$$

2.4 Solid state diffusion

In an electrode domain the diffusion of species into or out of a particle is described by Fick's law

$$\frac{\partial c_s}{\partial t} = \nabla(D_s \nabla c_s) \qquad (7)$$

in the r-dimension assuming ideal spherical particles. The boundary condition due to symmetry of the particle is given as

$$\frac{dc_s}{dr} = 0 \quad (\text{at } r = 0) \qquad (8)$$

at the particle's centre, whereas the boundary condition

$$-D_s \frac{\partial c_s}{\partial r} = j_+ \quad (\text{at } r_p = 0) \qquad (9)$$

accounts for the particle's surface. The right hand term describes the pore wall flux of Li/Li$^+$ at the electrode/electrolyte interface, which is positive for a de-intercalation of lithium. This corresponds to an anodic current (oxidation). Scaling with the porous electrode's specific surface and the volume fraction of the active material ε_s

$$a = 3\frac{\varepsilon_s}{r_p} \qquad (10)$$

allows for a macroscopic description of the pore wall flux. This is coupled with the mass balance in x-dimension by

$$R = aj_+. \qquad (11)$$

Since a single electron charge transfer mechanism

$$\Theta + n\text{Li}^+ + ne^- \leftrightarrows \text{Li}\Theta \qquad (12)$$

is assumed, the number of electrons n equals 1. In this case, Θ can be treated as an empty site in the porous electrode, which is graphite at the negative electrode (anode) and LiFePO$_4$ at the positive electrode (cathode). Thus, describes an intercalated Li$^+$.

To describe the conservation of charge within solid and liquid phase in an electrode domain

$$\nabla \mathbf{i}_l = Fnaj_+ + \mathbf{i}_{dl} \tag{13}$$

is given. To account for capacitive double layer effects the term $\mathbf{i}_{dl}$ is added. It describes the non-faradayic current due to a potential drop at the electrode/electrolyte interface. Since we are examining small signal frequency behaviour, adding this term is crucial. Whereas in steady-state conditions this term vanishes.

2.5 Reaction kinetics and double layer behaviour

The current-potential characteristics of an electrode are assumed to be following Butler-Volmer kinetics, so that

$$Fj_+ = i_{CT} = i_0 \left[\exp\left(\frac{\alpha_a F \eta}{RT} \right) - \exp\left(-\frac{\alpha_c F \eta}{RT} \right) \right] \tag{14}$$

where

$$i_0 = F(k_c)^{\alpha_a} (k_a)^{\alpha_c} (c_{s,max} - c_s)^{\alpha_a} (c_s)^{\alpha_c} \left(\frac{c_l}{c_{l,ref}} \right)^{\alpha_a} \tag{15}$$

is the exchange current density in A/m² which is defined by the electrode kinetic properties such as the charge transfer coefficient α (cathodic and anodic) and the concentration of charge at the electrode/electrolyte interface. In case of symmetric charge transfer coefficients the additional $c_{l,ref} = 1 mol / m^3$ reduces the reaction rate's unit k to m/s. The overpotential at the electrode/electrolyte interface is given by

$$\eta = \phi_s - \phi_l - E \tag{16}$$

with E being the equilibrium voltage of the electrode. Accounting for double layer behaviour an additional potential drop due to a non-faradayic current

$$\mathbf{i}_{dl} = \left(\frac{\partial \phi_s}{\partial t} - \frac{\partial \phi_l}{\partial t} \right) a C_{dl} \tag{17}$$

is implemented. For macroscopic analysis a further scaling with the electrode's specific surface and its specific double layer capacity C_{dl} in F/m² is needed.

2.6 Porous media

To account for effects given by porous structures an effective factorisation for a macroscopic view is necessary. Thus, transport parameters like diffusion coefficients and conductivities are scaled with the volume fraction of the related phase. Based on an empirical expression by Bruggeman [15] a sufficient approximation can be achieved by assuming

$$D_l = D_{l,0}(\varepsilon_l)^{1.5} \tag{18}$$

where $D_{l,0}$ represents the bulk diffusion coefficient. All volume fractions—active material, liquid and non-active—sum up to 1

$$\varepsilon_s + \varepsilon_l + \varepsilon_{non-active} = 1 \tag{19}$$

2.7 Boundary conditions

In order to give a reference for the potential, the anode side of this model is set to zero

$$\phi_s = 0 \quad (\text{at } x = 0). \tag{20}$$

On the cathode side the applied current density (in case of galvanostatic discharge/charge) is defined as

$$-\sigma \nabla \phi_s = s_{,app} \quad (\text{at } x = t_{neg} + t_{sep} + t_{pos}). \tag{21}$$

Since ions cannot leave the domains, the flux has to be set to zero at the boundaries

$$\mathbf{N}_+ = 0 \quad (\text{at } x = 0) \tag{22}$$

$$\mathbf{N}_+ = 0 \quad (\text{at } x = t_{neg} + t_{sep} + t_{pos}). \tag{23}$$

The potential within the liquid phase is defined as

$$\frac{d\phi_l}{dx} = 0 \quad (\text{at } x = 0) \tag{24}$$

$$\frac{d\phi_l}{dx} = 0 \quad (\text{at } x = t_{neg} + t_{sep} + t_{pos}). \tag{25}$$

2.8 Parameters

The model parameters are listed in Table 1.

2.9 Solving

The aforementioned equations are implemented in COMSOL Multiphysics. For the parameter variation the calculation was performed in the frequency domain. So the equations were transformed to frequency domain by applying FFT. For implementing a harmonic stimulation, the boundary condition (Eq. 21) is set to a sinusoidal signal. The impedance of the entire cell is given by

$$Z(\omega) = \frac{\phi(\omega)}{i_s(\omega)} \quad (\text{at } x = t_{neg} + t_{sep} + t_{pos}) \tag{26}$$

Frequencies between 1 mHz and 10 kHz are investigated. As the amplitude of stimulation a signal of 100 mA was chosen.

3 RESULT AND DISCUSSION

It should be noted that in the following diagrams a specific value of the cell's impedance is shown. It is related to the cell's active area, which is about 0,171 m² in this case. To conserve comparability to other values related to the active area, the values of impedance are listed as Ohm·m². Please note further that especially at high frequencies these values are very small and might be interpreted as zero, although they are not.

3.1 State of charge

Figure 1 shows the Bode plot for the State Of Charge (SOC) between 25% and 100%. Between 1 Hz and 1 mHz, the SOC accounts for over 100% change in the absolute value of

Table 1.   Model parameters.

Parameter	Value	Source
Geometry		
Cathode layer thickness t_{pos}	70 µm	[11]
Anode layer thickness t_{neg}	34 µm	[11]
Separator thickness t_{sep}	20 µm	Assumed
Volume fraction of solid phase $\varepsilon_{s,pos}$	0.43	[11]
Volume fraction of solid phase $\varepsilon_{s,neg}$	0.56	[11]
Volume fraction of electrolyte $\varepsilon_{l,pos}$	0.33	[11]
Volume fraction of electrolyte $\varepsilon_{l,neg}$	0.27	[11]
Volume fraction of electrolyte in separator $\varepsilon_{l,sep}$	0.5	[11]
Particle radius $r_{p,pos}$	36.5 nm	[11]
Particle radius $r_{p,neg}$	3.5 µm	[11]
Maximum concentration $c_{s,max,pos}$	22806 mol/m³	[11]
Maximum concentration $c_{s,max,neg}$	31370 mol/m³	[11]
Kinetics		
Equilibrium voltage E_{pos}	Analytic term	[11]
Equilibrium voltage E_{neg}	Analytic term	[11]
Reaction rate $k_{a,pos} = k_{c,pos} = k_{pos}$	0.3 e-11 m/s	Estimated
Reaction rate $k_{a,neg} = k_{c,neg} = k_{neg}$	0.35 e-11 m/s	Estimated
Transport		
Electronic conductivity $\sigma_{pos,0}$	10 S/m	Assumed
Electronic conductivity $\sigma_{neg,0}$	100 S/m	Assumed
Diffusion coefficient $D_{s,pos}$	1.18 e-18 m²/s	[11]
Diffusion coefficient $D_{s,neg}$	2 e-14 m²/s	[11]
Charge transfer coefficient $\alpha_{a,pos} = \alpha_{c,pos} = \alpha_{pos}$	0.5	Assumed
Charge transfer coefficient $\alpha_{a,neg} = \alpha_{c,neg} = \alpha_{neg}$	0.5	Assumed
Salt diffusivity in the electrolyte $D_{l,0}$	4 e-10 m²/s	[13]
Ionic conductivity of electrolyte κ_0	1 S/m	[13]
Activity dependency $\left(1 + \dfrac{d \ln f_{\pm}}{d \ln c_l}\right)$	1	Assumed
Transference number t_{+}	0.38	[13]
Double dlayer properties		
Specific capacity (cathode) $C_{dl,pos}$	0.2 F/m²	Est. based on [14]
Specific capacity (anode) $C_{dl,neg}$	0.2 F/m²	Est. based on [14]

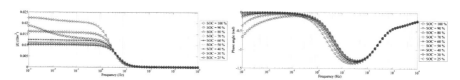

Figure 1.   Influence of the cell's state of charge (SOC) on impedance amplitude (left) and corresponding phase angle (right).

the impedance |Z| from 0.01 Ωm² to 0.023 Ωm² and phase shifts reaching from 0 rad to −0.7 rad at 1 mHz. However, the correlation of these changes with varying SOCs is rather ambiguous, as the plot in Figure 1 clearly reveals: 25% and 30% yield the two biggest absolute values, followed in the order of 100%, 40% and then 90% SOC.

3.2 Particle radius

The left side of Figure 2 shows the Bode plot for the cathode particle radius' variation at 100% SOC varies from 0 Ωm² to 0.026 Ωm² and the phase angle attains values from 0 rad to −1.3 rad. For frequencies of 1 Hz and above, the particle radius' influence on the spectrum is negligible. For lower frequencies, however, a certain influence can be observed: Towards lower frequencies, |Z| steadily grows with increasing particle radii. The phase shift between 2 mHz and 1 Hz is slightly more distinct for bigger particles. It nearly reaches zero in the region of 0.1 Hz.

The Bode plot for the cathode particle radius' variation at 20% SOC is shown in Figure 2 on the right. varies from 0 Ωm² to 0.014 Ωm², while the phase angle attains values from 0 rad to −1.3 rad. Unlike the case of 100% SOC, there is no distinct phase shift at low frequencies. Further, |Z| remains nearly constant in the low frequency region. For all frequencies, the particle size's influence can be treated as negligible. It might be stated that there is no dependency to the cathode's particle radius at low SOCs.

Figure 3 (left side) shows the anode particle radius' variation at 100% SOC varies from 0 Ωm² to 0.028 Ωm² and the phase angle from −1.3 rad to 0 rad. In contrast to the positive particle radius, the variation of the negative particle radius has a significant influence over the entire frequency range. The anode's particle radii are greater by two magnitudes in comparison to the cathode's particle radii. Hence, bounded diffusion should come into effect only for very low frequencies and semi-infinite conditions should prevail. It can be further observed that the phase shift in the high frequency domain grows with particle size.

At an SOC of 20%, nearly does not grow between 10 mHz and 1 Hz in comparison to 100% SOC. Still, bigger particles result in higher values of |Z|. A reduced dependency of the phase shift on particle radii variation can be observed in the low frequency range compared to 100% SOC. For frequencies above 1 Hz, the phase angle characteristics at 20% SOC coincide with those at 100% SOC. These results seem to underline the observation that the SOC mainly has an effect at low stimulation frequencies.

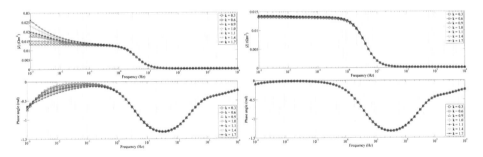

Figure 2.   The variation of the cathode's particle radius at a SOC of 100% (left) and 20% (right).

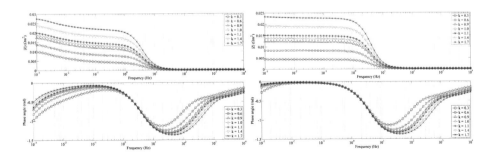

Figure 3.   The variation of the anode's particle radius $k \cdot r_{p,neg}$ at a SOC of 100% (left) and 20% (right).

3.3 Diffusion coefficients

Figure 4 (left side) shows different values of the diffusion coefficient on the cathode side at 100% SOC. In the low frequency region, slightly grows towards smaller frequencies, with higher growth rates for smaller particles. After reaching nearly zero at 0.1 Hz, the phase shift between 1 mHz and 1 Hz is slightly greater for smaller values of $D_{s,pos}$.

The effects of $D_{s,pos}$ on impedance behaviour are comparable to those of particle size, as higher diffusion coefficients result in a shift of bounded diffusion behaviour towards higher frequencies, as can be seen in Figure 4. At 20% SOC, it can be observed that varying $D_{s,pos}$ by several magnitudes nearly has no impact on the spectrum. This is particularly obvious with regard to the low frequency region, where distinct effects can be observed at 100% SOC. The results of a variation of the anode's diffusion coefficient $D_{s,neg}$ are shown in Figure 5. In contrast, the effects of varying $D_{s,neg}$ at 100% SOC are negligible and only observable at frequencies smaller than 0.1 Hz. Between 1 mHz and 10 mHz, $|Z|$ is slightly greater for smaller values of $D_{s,neg}$. At 20% SOC a drastic reduction of $D_{s,neg}$ has a clear effect at frequencies smaller than 0.1 Hz: $|Z|$ and the phase shift grows significantly.

3.4 Reaction rates

Figure 6 shows the Bode plot for different k_{pos} at 100% SOC and 20% SOC. Overall, little impact can be observed. Independent of SOC, $|Z|$ is slightly greater for lower k_{pos} at low frequencies.

The anode's reaction rate k_{neg} has a much more obvious influence on the spectrum than k_{pos}. Figure 7 shows the Bode plot for 100% SOC and 20% SOC. Here, $|Z|$ reaches values up to nearly 0.05 Ωm^2 in the low frequency region. Smaller values of k_{neg} result in higher absolute values of $|Z|$. However, the general tendency of $|Z|$ being rather constant and the phase shift being close to zero, still applies between 1 mHz to 1 Hz. The phase angle also changes as follows: At 0.1 Hz and below, smaller values of k_{neg} result in greater phase shifts. Between

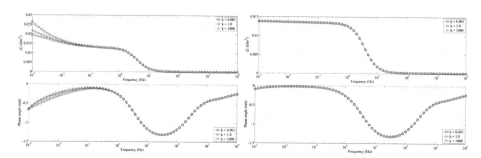

Figure 4.   The variation of the cathode's diffusion coefficient $k \cdot D_{s,pos}$ at a SOC of 100% (left) and 20% (right).

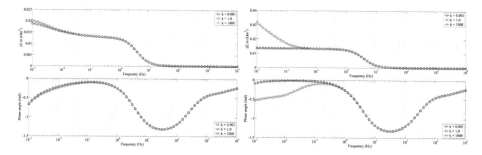

Figure 5.   The variation of the anode's diffusion coefficient $k \cdot D_{s,neg}$ at a SOC of 100% (left) and 20% (right).

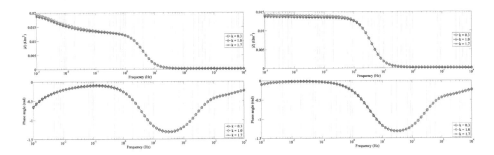

Figure 6. The variation of the cathode's reaction rate $k \cdot k_{pos}$ at a SOC of 100% (left) and 20% (right).

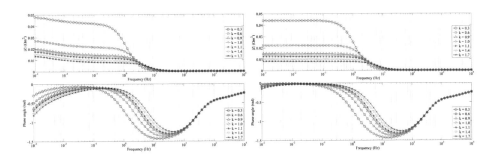

Figure 7. The variation of the anode's reaction rate $k \cdot k_{neg}$ at a SOC of 100% (left) and 20% (right).

0.1 Hz and 100 Hz, however, this behaviour is reversed. At 20% SOC, a variation of shows similar effects, whereas a slight variation k_{neg} below 10 mHz can be noticed.

3.5 Conclusion

Within the research work presented here, it could be shown that a physico-chemical model of a Li-ion battery is capable of evaluating the influence of various cell parameters and material properties on the cell's impedance characteristics. Whereas the influence of the cell's SOC on the impedance characteristics can be easily determined via experiment and therefore validated with little effort, more fundamental properties of the active materials and the electrolyte such as particle radius, diffusion coefficients and reaction rate coefficients can be evaluated, as well. Although quite obvious effects of varying certain cell characteristics can be observed via simulation, some of the results presented here seem to be ambiguous with respect to experimental studies. In general terms it can be stated that the SOC, the anode's reaction rate coefficient k_{neg} and the anode's particle radius $r_{p,neg}$ have the greatest influence on the spectrum. The fact that $r_{p,pos}$ and k_{pos} have little influence suggests that the anode is dominating. Surprisingly, the diffusion coefficients of both cathode and anode only produce minor changes in the spectrum even when varied over 6 magnitudes. Varying the SOC produces a characteristic effect, which superimposes the influence of other parameters when varied simultaneously. Therefore, more research work has to be performed in terms of improving the model parameters and validation of simulation results. This is part of the following research activities.

ACKNOWLEDGEMENT

This work was supported by the German Federal Ministry of Education and Research under contract 03X4627A (project name: LiSSi).

REFERENCES

[1] M.E. Orazem and B. Tribollet, "An integrated approach to electrochemical impedance spectros-copy," *Electrochimica Acta*, vol. 53, no. 25, pp. 7360–7366, 2008.

[2] G. Sikha and R.E. White, "Analytical expression for the impedance response for a lithium-ion cell," *Journal of the Electrochemical Society*, vol. 155, no. 12, A893–A902, 2008.

[3] J.R. Macdonald and E. Barsoukov, "Impedance Spectroscopy: Theory, Experiment, and Applica-tions," *History*, vol. 1, p. 8, 2005.

[4] M. Doyle, J.P. Meyers, and J. Newman, "Computer simulations of the impedance response of lithium rechargeable batteries," *Journal of the Electrochemical Society*, vol. 147, no. 1, pp. 99–110, 2000.

[5] W.G. Bessler, "A new computational approach for SOFC impedance from detailed electrochemical reaction–diffusion models," *Solid State Ionics*, vol. 176, no. 11, pp. 997–1011, 2005.

[6] D.W. Dees, K.G. Gallagher, D.P. Abraham, *et al.*, "Electrochemical Modeling the Impedance of a Lithium-Ion Positive Electrode Single Particle," *Journal of the Electrochemical Society*, vol. 160, no. 3, A478–A486, 2013.

[7] A. Häffelin, J. Joos, M. Ender, *et al.*, "Time-Dependent 3D Impedance Model of Mixed-Conducting Solid Oxide Fuel Cell Cathodes," *Journal of the Electrochemical Society*, vol. 160, no. 8, F867–F876, 2013.

[8] M. Doyle, T.F. Fuller, and J. Newman, "Modeling of galvanostatic charge and discharge of the lith-ium/polymer/insertion cell," *Journal of the Electrochemical Society*, vol. 140, no. 6, pp. 1526–1533, 1993.

[9] T.F. Fuller, M. Doyle, and J. Newman, "Simulation and optimization of the dual lithium ion inser-tion cell," *Journal of the Electrochemical Society*, vol. 141, no. 1, pp. 1–10, 1994.

[10] J. Newman and K.E. Thomas-Alyea, *Electrochemical Systems*. John Wiley & Sons, 2012.

[11] M. Safari and C. Delacourt, "Modeling of a commercial graphite/LiFePO$_4$ cell," *Journal of the Electrochemical Society*, vol. 158, no. 5, A562–A571, 2011.

[12] S.-Y. Chung, J.T. Bloking, and Y.-M. Chiang, "Electronically conductive phospho-olivines as lith-ium storage electrodes," *Nature materials*, vol. 1, no. 2, pp. 123–128, 2002.

[13] L.O. Valøen and J.N. Reimers, "Transport properties of LiPF$_6$-based Li-ion battery electrolytes," *Journal of the Electrochemical Society*, vol. 152, no. 5, A882–A891, 2005.

[14] S. Brown, N. Mellgren, M. Vynnycky, *et al.*, "Impedance as a Tool for Investigating Aging in Lith-ium-Ion Porous Electrodes. II. Positive Electrode Examination," *Journal of the Electrochemical Society*, vol. 155, no. 4, A320–A338, 2008.

[15] V.D. Bruggeman, "Berechnung verschiedener physikalischer Konstanten von heterogenen Substan-zen. I. Dielektrizitätskonstanten und Leitfähigkeiten der Mischkörper aus isotropen Substanzen," *Annalen der physik*, vol. 416, no. 7, pp. 636–664, 1935.

Sensors

Lecture Notes on Impedance Spectroscopy, Volume 5 – Kanoun (Ed.)
© *2015 Taylor & Francis Group, London, ISBN 978-1-138-02754-1*

3D eddy current modelling for non destructive crack detection

Tino Morgenstern
Institute of Measurement Engineering and Sensor Technology, University of Applied Sciences HRW, Mülheim an der Ruhr, Germany

R. Brodskiy
ifm Electronic GmbH, Tettnang, Germany

J. Weidenmuller
Institute of Measurement Engineering and Sensor Technology, University of Applied Sciences HRW, Mülheim an der Ruhr, Germany

O. Kanoun
Faculty for Electrical Engineering and Information Technology, Technische Universität Chemnitz, Chemnitz, Germany

J. Himmel
Institute of Measurement Engineering and Sensor Technology, University of Applied Sciences HRW, Mülheim an der Ruhr, Germany

ABSTRACT: Eddy current sensors are well-established in industrial applications, such as in hot testing of rolled rods. However, the eddy current sensors currently in use are not able to detect longitudinal cracks in hod rolled rods. This paper introduces a 3D eddy current model which allows an investigation of this particular problem. Generally, 3D eddy current Finite Element Analysis (FEA) are time consuming due to the fine mesh as a result of the small penetration depth of the currents. The introduced 3D FEA model calculates the complex coil impedance of a detection coil affected by a rod with cracks using the current sheet approximation with additional correction formulas.

Keywords: Eddy Current Modelling; Crack Detection

1   INTRODUCTION

The finite element method is already established in modelling for eddy current sensors the same as analytical methods. For example, reference [1] solves for the absolute impedance of core-less coils, and [2], [3] solve for the induced voltage in the coil. The problem with these analytical models is the dependency on the crack size, shape and location, and the equations needs to be modified all the time. Consequently, the results may not necessarily be comparable. When the FEA is used instead, the results remain comparable for changed model parameters. Hence, the aim of this study is to build a 3D model to calculate the complex coil impedance for encircling coils that measure rods with any kind of cracks. In doing so, two main simplifications are applied as follows.

First, the impedance boundary conditions [4] for an improved modelling of small skin depth in the rod is used.

Second, the current sheet approximation is used (Multi Turn Coil Domain) [5], which reduces the total number of elements by summing up all windings into one sheet. This measure, in turn, requires some further correction to estimate the neglected inter-winding capacitance and coil wire length. This correction is performed using complex material properties which can be used in 2D/axisymmetric finite element models [6]. Here these complex material properties were used to model a 3D coil surrounding a target. For improving the applicability of this two main simplifications preliminary investigations were done in 2D with the advantages of reduced number of elements, thus a shortened calculation times. A validation of both 2D and 3D models will be given in the result chapter.

2 MATERIALS AND METHODS

This chapter focuses on the modelling aspects and physics of the 3D FEA. The building of the FEA model is based on several topics.

First, the adequate physics (e.g. methods and formulas) needs to be chosen which is realised in 2D models. In this case, the magnetic field interface (mf) was selected, which solves for the vector potential using the following equation [5]:

$$\nabla \times (\frac{1}{\mu_r \mu_0} \nabla \times \mathbf{A}) + j\omega\sigma\mathbf{A} - \omega^2 \varepsilon_0 \varepsilon_r \mathbf{A} = \mathbf{J}_e \qquad (1)$$

where: μ_0 = permeability in free space, μ_r = relative permeability, ε_0 = permittivity in free space, ε_r = relative permittivity, ω = angular frequency, j = complex unit, σ = conductivity, A = magnetic vector potential and $\mathbf{J}_e$ = external current density defined by the user (see Equation 3). Using the adequate physics it is possible to transform the 2D into a 3D geometry. Since the crack is not symmetrical, it is necessary to work with a 3D model, which consists of a Cartesian coordinate system.

Second, all geometries are modelled in 3D and the corresponding domain and boundary conditions are applied, as stated in Table 1.

Figure 1 shows a complete model, including all domains, boundaries and the reference edge for the Multi Turn Coil Domain. The coil parameters are, the inner coil radius r_{icoil} = 12.5 mm, the coil length l_{coil} = 25 mm and the coil wire radius r_{wire} = 0.5 mm.

Table 1. Domains, boundaries and material properties.

		Material properties	
Type	Material	μ_r	σ
Domain Ω_1	air	1	1S/m
Boundary Γ_1	air	1	1S/m
Domain Ω_2	copper	μ_{eff}	σ_{eff}
Boundary Γ_3	brass	1	13.7 MS/m

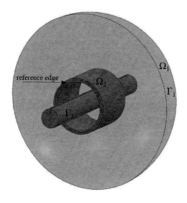

Figure 1. Drawing of all domains and boundaries.

Within the typical frequency range, between 0.5 MHz and 1 MHz for eddy current measurements, the skin depth δ lies between 190 μm and 140 μm, which is much smaller then the rod diameter, $d = 9$ mm. In order model such a small skin depth a boundary layer mesh becomes necessary which create 8 elements on 390 μm with element sizes between 22–90 μm. Alternative an important feature of the model is the use of the impedance boundary condition [4]. The penetration depth of the electric (**E**) and magnetic (**H**) fields into the rod is approximated using Equation 2:

$$\sqrt{\frac{\mu_0 \mu_r}{\varepsilon_0 \varepsilon_r - j\sigma/\omega}} \mathbf{n} \times \mathbf{H} + \mathbf{E} - (\mathbf{n} \cdot \mathbf{E})\mathbf{n} = 0 \qquad (2)$$

As a result, the inner part of the rod does not need to be solved and remains mesh-free and information of the current densities is needed. This measure, in turn, positively affects the total number of elements.

Here the model consisting of a rod with a diameter $d = 9$ mm and a length $l = 400$ mm and a coil using the current sheet approximation in 3D. This leads to following number of elements. Applying the impedance boundary condition reduces the number of elements to 426,000 elements instead of 1,310,000 elements. Modelling the rod using impedance boundary condition gets realised with a boundary mesh. The number of this boundary mesh elements results of surrounding volume mesh elements.

The second interesting aspect in the model is the use of the current sheet approximation, which serves as a fundamental aspect in the Multi Turn Coil Domain [5]. The current I, which flows through the coil, is set, and the external current density J_e is calculated with the following equation.

$$J_e = \frac{\omega \cdot I}{A_{coil}} \cdot e_{wire} \qquad (3)$$

where: ω = number of turns, A_{coil} = coil cross section area and e_{wire} = unit direction vector of the wire.

Additional parameters for defining the Multi Turn Coil Domain are shown in Table 3.

Unfortunately, all side effects, such as the frequency dependent wire resistance and inter-winding capacitances, are neglected in the Multi Turn Coil Domain, hence making further correction formulas necessary. These side effects are approximated by implementing complex material properties to calculate the frequency dependent resistivity ρ_r in 5 for the coil [6]:

Table 2. Overview: Number of volume elements.

	Impedance boundary coditions	Boundary layer mesh
Model without rod	426000	426000
rod	0	884000
Model with rod	426000	1310000

Table 3. Multi turn coil domain definitions.

Parameter	Value
coil type	circular
reference edge	mean coil radius
excitation current	1A
number of turns	21

Table 4. Overview coil modelling.

	Single windings	Current sheet
Number of Elements (2D)	168302	33776
Number of Elements (3D)	$1.2 \cdot 10^6$	$5.9 \cdot 10^4$

$$F = \frac{\sigma \mu_0 \omega}{2} \cdot r^2 \tag{4}$$

$$\rho_r = \left(\frac{1}{\sigma \cdot fill} \right) \left(\frac{\sqrt{jc_3 F}}{tanh(jc_3 F)} + jc_4 F \right) \tag{5}$$

$$\mu_{eff} = (1 - c_2)\mu_0 + c_2 \mu_0 \cdot \frac{tanh(\sqrt{jc_1 F})}{\sqrt{jc_1 F}} \tag{6}$$

The complex permeability μ_{eff} is calculated by means of the non-dimensional frequency F and the functions c_1 and c_2. The functions, c_1 to c_4, are functions of the coils fill factor $fill$.

$$c_2 = \frac{3}{2} \frac{fill}{c_1} \tag{7}$$

$$c_1 = -0.0714373 fill^3 + 0.0684158 fill^2 + 0.687385 fill + 0.775607 \tag{8}$$

$$c_4 = 0.425218 - \frac{log(fill)}{2} - \frac{c_3}{3} \tag{9}$$

$$c_3 = -0.215718 fill^3 + 0.722321 fill^2 - 0.00860551 fill + 0.882464 \tag{10}$$

In Equations 8 and 10, c_1 and c_3 are approximated cubic polynomial functions depending on coils fill factor for using the complex material properties in 2D FEA. The functions c_3 and c_4 are necessary to calculate ρ_r in Equation 5. [6]

$$\sigma_{eff} = \frac{1}{\rho_r - \frac{1}{3} j \omega \mu_{eff} (b + \varepsilon)^2} \tag{11}$$

In Equation 11, ρ_r is the frequency-dependent resistivity, b performs with the coil wire radius and ε describe the distance between two layers in multi-layer coils.

Using the current sheet approximation with complex material properties it is possible to reduce the total element numbers of the subdomain coil (Table 4).

3 RESULTS AND DISCUSSION

The simulated results of 2D models are shown in Figure 2, where the following configurations are under investigation:

- Setup 1: coil in air (length: 25 mm; diameter: 25 mm; windings: 21), with 21 single windings
- Setup 2: coil (Setup 1), using the current sheet approximation without complex material properties
- Setup 3: coil (Setup 2), using complex material properties.

Following the simulated results are compared with laboratory measurements.

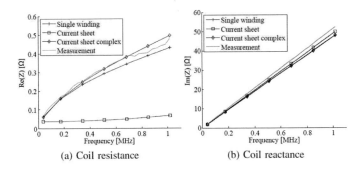

(a) Coil resistance   (b) Coil reactance

Figure 2.   Results of 2D simulation models.

Table 5.   Deviation at 1 MHz.

	Deviation at 1 MHz [%]	
	Re	Im
Single windings	10,17	8,43
Current sheet	85,72	5,05
Current sheet compl	3,48	8,45

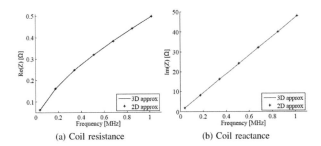

(a) Coil resistance   (b) Coil reactance

Figure 3.   Results of 2D and 3D simulation.

The deviation of the simulation results in 2D are shown in Table 5. Here (Setup 3) with the current sheet approximation shows the best results in both, the resistance and reactance. Evaluation the purely inductive region the reactance of Setup 2 shows the best results. Here the disadvantage is the huge deviation in the resistance with 85%. Reasons for this result are the neglected frequency-dependent wire resistance resulting from the skin depth and inter-winding capacitances in the Multi Turn Coil Domain.

In the following the 2D and 3D simulations using the current sheet approximation with complex material properties became compared. Viewing the resistive part the maximum deviation amounts 0.16% at 511 kHz. However, the reactance shows the maximum deviation at 1 MHz with 0.01%. So it is possible to use the complex material properties including the approximated functions c_1 to c_4 for 3D finite element analyses. This 2D and 3D eddy current modelling using the current sheet approximation shows good results.

All simulated (3D) and measured results with crack influence are shown in Figure 4, where the following configurations are under investigation:

- Setup 4: coil (length: 25 mm; diameter: 25 mm; windings: 21) measures a perfect rod
- Setup 5: coil (Setup 4) measures a rod with a crack (depth = 1 mm, width = 0.6 mm).

A few values from the study are compared in Table 6, where $\Delta R = R'(crack) - R(nocrack)$ and $\Delta X_L = X'_L(crack) - X_L(nocrack)$. The Value Δ represents the change of the resistance caused by the crack and ΔX_L represent the change of the reactance caused by the crack.

69

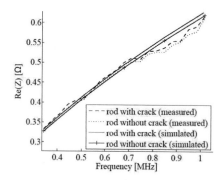

Figure 4. Simulation results in comparison to laboratory measurements.

Table 6. Crack influence.

Crack influence	Frequencies [MHz]			
	0.5	0.68	0.85	1
Simulation $\frac{\Delta R}{R}[\%]$	1.62	1.62	1.58	1.55
Measurements $\frac{\Delta R}{R}[\%]$	1.0	0.97	1.99	0.56
Simulation $\frac{\Delta X_L}{X_L}[\%]$	0.09	0.10	0.14	0.12
Measurements $\frac{\Delta X_L}{X_L}[\%]$	0.04	0.03	0.03	0.02

In general, all simulations delivered good results as compared to the measurements. However, the crack influence on the measurements seems to be rather small and, hence the coil characteristics needs to be further adjusted for the crack detection application.

4   CONCLUSION

In conclusion, the model performs well within the purely inductive frequency range. Both simplifications, the impedance boundary condition on the rod and the current sheet approximation with complex material properties, were successfully applied to the FEA model in 2D as well in 3D. As a result, the total element number was reduced by a factor of 3. Consequently, the simulation model allows extensive parameter studies for several types of cracks measured with different encircling coils.

ACKNOWLEDGMENTS

Thanks to Transfer NRW Ziel2 FH-Extra which is funding this research project (In-Situ-Messung der Querschnittsfläche von Warmwalzhalbzeugen mit integrierter Risserkennung) from 1/2011 up to 12/2013.

REFERENCES

[1] F. Wendler, U. Tröltzsch, and O. Kanoun, "Modellierung der absoluten Impedanz einer Luftspule mit Wirbelstromrückwirkung," in *Technisches Messen*, vol. 79, 2012, pp. 516–521.
[2] A. Fuchs, W. Häuselhofer, and G. Brasseur, "Design of an eddy current based crack detection sensor for wire processing applications," in *IEEE Sensors conference*, 2004, pp. 840–843.

[3] M. Rahman and R. Markelin, "Numerische Modellierung und Optimierung von Wirbelstromsensoren zur Online-Heidrahtprüfung," in *DGZfP-Jahrestagung*, 2009, pp. 1–8.

[4] T. Senior, "Impedance boundary conditions for imperfectly conducting surfaces," *Applied Scientific Research, Section B*, vol. 8, no. 1, pp. 418–436, 1960.

[5] *AC/DC Module Users Guide*, Version 4.3., Comsol, 2012.

[6] D.C. Meeker, "An improved continuum skin and proximity effect model for hexagonally packed wires," *Journal of Computational and Applied Mathematics*, vol. 236, no. 18, pp. 4635–4644, 2012.

Lecture Notes on Impedance Spectroscopy, Volume 5 – Kanoun (Ed.)
© *2015 Taylor & Francis Group, London, ISBN 978-1-138-02754-1*

Investigation of the sensitivity profile of an analytical eddy current model

Frank Wendler, Paul Büschel & Olfa Kanoun
Chair of Measurement and Sensor Technology, Technische Universität Chemnitz, Chemnitz, Germany

ABSTRACT: In this contribution the influences of the main model parameters of eddy current problems are investigated. The sensitivity of distance, conductivity and permeability on the inductance are computed and analyzed for a frequency range from 40 Hz to 100 MHz. The model is based on a core-less coil with 10 turns, a diameter of 11 mm and a target with properties similar to steel. The aim of this investigation is to identify sensitivity patterns and analyze their response to large variations of the model parameters for the purpose of hardware design and computing the inverse problem. The results represent the general characteristics of core-less eddy current sensors at large variations of those parameters and are discussed in a general way.

Keywords: eddy current; core-less coil; sensitivity; inductance

1 INTRODUCTION

Eddy current sensors are one of the most common distance sensors in industrial applications. The robust measurement concept and the encapsulated design ensure long term operation under harsh conditions like heat, dust, radiation and corrosive media. A common approach is a single frequency measurement and a comparison to the characteristic distance function. By using a single frequency analysis only one parameter can be obtained. Other parameters like conductivity and permeability need to be considered as constant. For this reason the change of the characteristic distance function with changing properties of the targeted material requires a calibration process in most applications. Another approach is to increase the number of parameters obtained with one measurement. This requires the increase the measured frequency points to collect the information to separate the additional parameters. For this spectroscopic method, a broad band analysis of the sensitivities for each model parameter is helpful in hardware design and for developing strategies for data analysis. The aim of this contribution is to detect frequency regions of high sensitivity and unique sensitivity profiles, which can be used for a separation and estimation of multiple parameters. For this investigation an analytical coil model is derived numerically to obtain the sensitivities over three decades of each parameter of the model.

2 THE EDDY CURRENT MULTI-TURN MODEL

The model was obtained from magnetic field models in [1] for single turn coils. This model is a reduced form of the model of Dood and Deeds [2]. Only the distance to the surface of a flat homogeneous object with a certain conductivity and magnetic permeability are parameters of this model. The resulting magnetic vector potential for the use in multi-turn coil applications, as shown in figure 1, can be obtained by the superposition of the vector potentials of each single turn:

$$A_{total} = A_1 + A_2 + \cdots + A_N. \tag{1}$$

The inductance can be calculated by integrating the vector potential over path of the electric current:

$$L = \frac{Z}{i\omega} = \frac{U}{i\omega I} = \frac{i\omega \oint A \, ds}{i\omega I} = \frac{1}{I} \oint A \, ds. \tag{2}$$

By using the superimposed vector potential in equation 2, equation 3 can be obtained for the first turn:

$$L_1 = A_{total,1} \cdot \frac{2\pi r_1}{I_1} = \underbrace{A_{1,1} \cdot \frac{2\pi r_1}{I_1}}_{L_{1,1}} + \cdots + \underbrace{A_{N,1} \cdot \frac{2\pi r_1}{I_1}}_{L_{N,1}}. \tag{3}$$

The addends of equation 3 can be written in matrix form:

$$\begin{pmatrix} L_1 \\ L_2 \\ \vdots \\ L_N \end{pmatrix} = \begin{pmatrix} L_{1,1} & L_{2,1} & \cdots & L_{N,1} \\ L_{1,2} & L_{2,2} & \cdots & L_{N,2} \\ \vdots & \vdots & \ddots & \vdots \\ L_{1,N} & L_{2,N} & \cdots & L_{N,N} \end{pmatrix}. \tag{4}$$

As the sending and receiving turns of the coil are the same for the elements on the main diagonal of this matrix, these elements are self-inductances. An integration path on the surface of the wire has to be used for those self-inductances to apply the valid boundary values [3], [4]. For example the first element on the main diagonal is equation 5 [5], [6]:

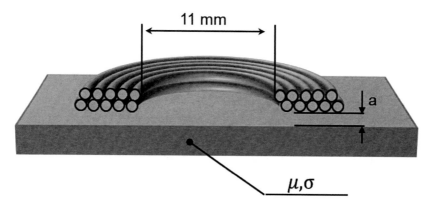

Figure 1.   Coil above the target material (not drawn to scale).

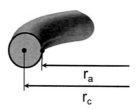

Figure 2.   Integration path on the wire surface at r_a.

74

$$L_{1,1} = \mu_o \pi r_{a_1} \int_0^\infty J_1(k) J_1\left(\frac{kr_{a1}}{r_{c1}}\right)\left[1 + \frac{k\mu_r - \sqrt{k^2 + i\omega\sigma\mu r_{c1}^2}}{k\mu_r + \sqrt{k^2 + i\omega\sigma\mu r_{c1}^2}} e^{\frac{k(-2a)}{r_{c1}}}\right] dk \qquad (5)$$

The Inductance of the entire coil is the sum of all self and coupling inductances of the individual coils and can be calculated as:

$$L_{total} = \sum_{n=1}^{N} \sum_{m=1}^{N} L_{n,m}. \qquad (6)$$

3 SENSITIVITY CALCULATION

3.1 Numerical perturbation

Equation 7 is used for calculating the relative change of the inductance due a symmetric change of the investigated parameter of 0,1%. Like illustrated in figure 3, each of the tree inductance values L_u, L_l and L_o is obtained by using equation 6 with 99.9%, 100% and 100.1% of the corresponding parameter value. The calculated sensitivity was normalized by the inductance without target influence L_o to provide a quantity without physical dimension for better comparability. The numerical integration of the analytical model was calculated using trapezoidal integration in Matlab varying k from 0 to 500 in a step size of 1.

$$S = \frac{(L_u - L_l) \cdot x_c}{(x_u - x_l) \cdot L_0} = \frac{L_u - L_l}{0,002 \cdot L_0} \qquad (7)$$

3.2 Geometric parameters

Equation 6 was used to calculate the sensitivity of a coil with the diameter of 11 mm and an wire diameter of 0.1 mm. The 10 turns were arranged a 2×5 wiring scheme in figure 1. The coil diameter is a medium diameter for eddy current distance sensors. The number of turns is chosen to provide a high inductance and resonance frequency above 10 MHz, similar to other eddy current probe experiments [7]. The wire diameter and the wiring scheme provide an compact coil, which can be manufactured using self bonding wire.

3.3 Parameter range

To investigate strategies for spectral data processing, the parameters need to cover a wide range and represent realistic combinations of parameters. For this analysis each of the three parameters has been varied over 3 decades in logarithmic spacing, to support the comparability of the results by similar plots. While the investigated parameter has been swept, all other model parameters were applied as standard conditions according to table 1. Those standard

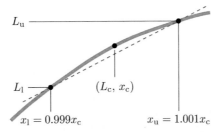

Figure 3. Sensitivity calculated as normalized numerical derivative—L_u upper value, L_l lower value, L_c corresponding value.

Table 1. Standard values of non varied parameters.

Target related values		Coil related values	
Conductivity σ	100 MS/m	Inner diameter	11 mm
Permeability μ	100	Wire diameter	100 μm
Distance *a*	1 mm	Scheme	2×5

conditions represent an area in the center of the parameter range and can be considered close to low permeable steel material at an effective working distance. The range of conductivity covers from 100 MS/m to 0.1 MS/m all metallic materials. The range of permeability from 1 to 1000 covers most common ferromagnetic materials. Materials for electro-magnetic and power applications may exceed this parameter range. The range of distance was chosen from 0.1 mm to 100 mm, which can be considered from a very close distance to far beyond the detection limit for this 11 mm sensor coil. The frequency range from 40 Hz to 100 MHz covers the often used frequency range for distance measurement from 1 kHz to 10 MHz and is extended to the frequency range of accessible measurement to support future investigations. The conductivity and the permeability are considered constant over frequency. Most publications show an extended plateau for singular, ferromagnetic materials over several decades of frequency with strong drop of the permeability at the end [8]–[11]. These spectra are made for different materials and vary strong in their observed frequency limit. Some investigations consider the magnetic field rejection by eddy currents and the influence of resonant effect of the experimental setup not to the desired extend. Other publication concerning the ferromagnetic resonance report a higher frequency limit of permeability between 1 GHz and 10 GHz, caused by the dynamic of domain wall motion [12]–[14]. Since no other permeability reducing mechanism of ferromagnetic materials are known to the author jet, the magnetic permeability is considered as stable, real value up to the frequency mentioned above.

4   RESULTS AND DISCUSSION

4.1   *Conductivity*

The rejection of magnetic fields out of the target is decreasing the coil inductance in a certain frequency range (figure 4a, dark area). At this frequency a negative real part oft the sensitivity coefficient is observed. The peak level of sensitivity is constant, but decreasing in frequency with increasing conductivity. In the plot of the imaginary part of the sensitivity in figure 4b a similar frequency shift with increasing conductivity can be observed. The sensitivity pattern itself is different. This pattern consists of an area of negative sensitivity, an area of positive frequency and an area of very low sensitivity in between. This area of low sensitivity is caused by a maximum of the imaginary part of the inductance at this frequency and results in a local weak response of the imaginary part of the inductance to small variations of conductivity.

4.2   *Permeability*

The rejection of the magnetic fields out of the target is also accruing in combination with high permeability materials. Contrary to the influence of conductivity, rising values of the magnetic permeability are shifting the inductance drop to higher frequencies. This is resulting in an increase of the sensitive frequency range of the real part in figure 5a with increasing permeability. The peak level of sensitivity is constant and has the same absolute value like in figure 4a, because of the modification of the same effect in the opposite way. At low values of permeability and low frequencies a strong positive sensitivity coefficient (bright area) can be observed in figure 5a. This additional effect is caused by the reduced magnetic resistance trough the target material at increasing values of permeability. The sensitivity to this effect is decreasing to with increasing permeability, because of the limiting contribution of the air

76

volume to the decrease of the magnetic resistance for the entire loop of the magnetic flux. The use of a core might amplify this magneto-static effect and extend the usable frequency range, but it will add a strong frequency dependency by the own magnetic properties of the core. The imaginary sensitivity plot in figure 5b shows the same sensitivity pattern from figure 4b with inverted sign and a shift of the pattern to higher frequencies with increasing permeability like in the plot of the real part in figure 5a. The contribution of the permeability to

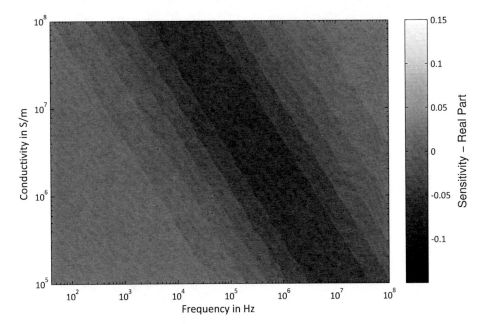

(a) Frequency shift of the sensitivity pattern from 10 MHz to 10 kHz (real part)

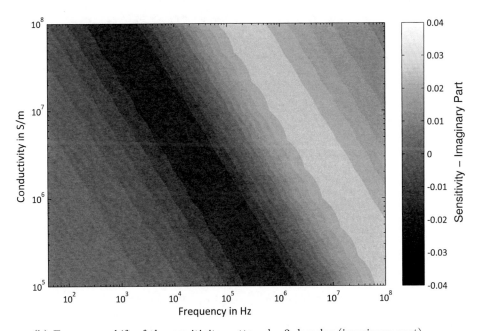

(b) Frequency shift of the sensitivity pattern by 3 decades (imaginary part)

Figure 4.   Sensitivity of the inductance to conductivity and frequency.

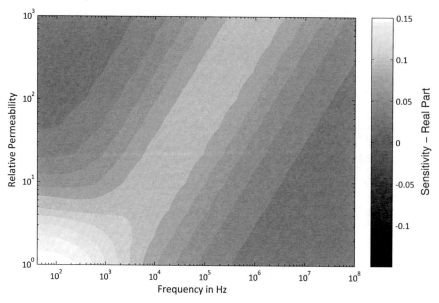

(a) Frequency shift of the sensitivity pattern from 1 kHz to 1 MHz and static magnetic contribution for low permeability (real part)

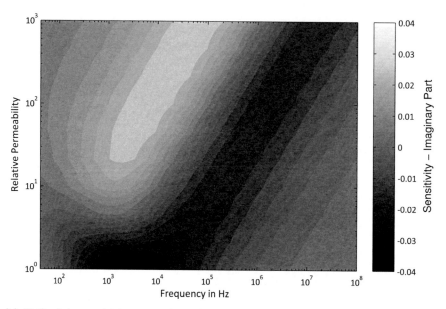

(b) Shift of the sensitivity patter by 3 decades and static magnetic contribution for low permeability (imaginary part)

Figure 5. Sensitivity of the inductance to permeability and frequency.

the imaginary part of the sensitivity at low frequencies is also increased by the static magnetic effect in the region of permeability from 1 to 7 (black area). This region is covers approximately half of the permeability range compared to the real part in figure 5a, but is located about 1 decade higher in frequency.

4.3 *Distance*

At low frequencies a high permeability value does increase coil inductance and at high frequencies the coil inductance will be decreased by the eddy current effect. Both regions will be affected by the distance characteristic and result in the respective real part of sensitivity coefficients in figure 6a. In between those regions exists a transition area were the real part of the coil inductance has the same value as a coil without target. If the inductance of the coil is unchanged by entirely removing the target, a scaling effect at this frequency cannot

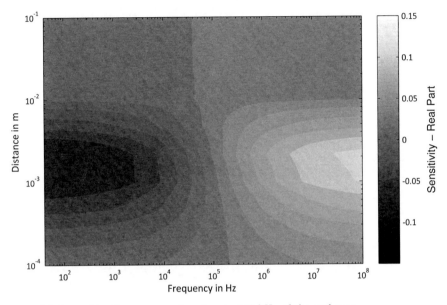

(a) Insensitive frequency region close to 100 kHz of the real part

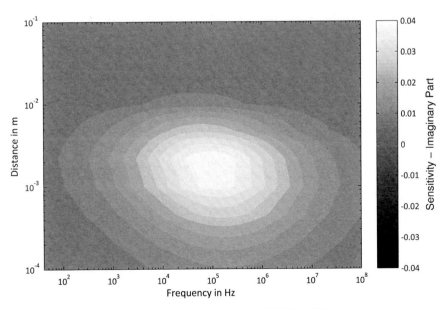

(b) Highly sensitive frequency region close to 100 kHz of the imaginary part

Figure 6.   Sensitivity of the inductance to target distance and frequency.

be observed by removing the target in distance steps either. The low sensitivity at distances above 10 mm are caused by the general limit of detection for eddy current sensors, which is approximately the diameter of the coil. In the plot of the imaginary part of the sensitivity in figure 6b a similar explanation can be applied. With the maximum of imaginary part of the inductance located in this transition region at 100 kHz, a scaling of this maximum by a change in distance does result in a strong sensitivity in this region. This region of sensitivity is limited by the coil diameter, like the real part of sensitivity. For the real and imaginary a low sensitivity for small target distances can be observed. Since the material dependent part in equation 1 is scaled exponentially with distance, a high sensitivity is expected at small distances. The difference between result and expectation might be a computational error caused by the limited step size or limited number of steps in the numerical integration and will be subject to further investigations in the future.

5   CONCLUSION

The influence of distance was identified as a scaling behavior without any shift in frequency. The influence of conductivity resulted in an shift of the sensitivity pattern without any scaling or change of shape of the pattern. The influence of permeability can distinguished in a magneto-static contribution and a influence on the eddy current effect. The magneto-static contribution can be observed at low frequencies and low permeability and can be described as nonlinear scaling behavior by the distance and the permeability. The influence on the eddy current effect is a frequency shift and opposing the influence of conductivity. The theoretical investigation furthermore implies that inverse eddy current problems have to be distinguished in low and high permeability applications and require different approaches. Based on the presented results several strategies for signal processing can be derived. One is to use a common approach in taking the advantage of analyzing the imaginary part of the inductance. The strong sensitivity in figure 6b is related to a distance sensitive maximum of the imaginary part in this frequency range, caused by the rejection of the magnetic field out of the target. In figure 4b and 5b the sensitive corridor at the same frequency range indicates, that small variation of the permeability or conductivity will not have much influence on this effect. This explains the good performance of single frequency techniques using the quality factor of the coil in combination with calibration to the target material. The insensitive region for variations of conductivity and permeability might not cover changes of the target material, but it will decrease the effects of small variation, caused by environmental conditions or limited reproducibility of the material parameters of the target. For a multiple frequency approach in spectral distance measurement this shifting effect can be used to adept to the material properties by tracking the frequency shift and obtaining the distance information from the value of this maximum. The similar pattern can be identified in the real pat of the sensitivity plot. The higher sensitivity of the real part of inductance implies a higher suitability for signal processing. The wide area insensitive to distance in the center of figure 6a is located in the same frequency range like the sensitive areas in figure 4a and 5a and provides information by the frequency shift. In this case, the absolute value of inductance in the sensitive regions in figure 6b provide the distance information. The additional effect in figure 5a and 6b for an increase in permeability in the permeability range of 1 to 10 is extending the sensitivity pattern and might be used to separate the material properties conductivity and permeability. In this range a separation of conductivity an permeability might be performed with the use of an eddy current probe. To cover the entire sensitivity pattern approximately 3 decades of the inductance spectra are needed. To track the frequency shift an additional decade of the inductance spectra is necessary for each decade of permeability or conductivity the system should cover.

REFERENCES

[1] J.-A. Tegopoulos and E.-E. Kriezis, *Eddy currents in linear conducting media*. Elsevier, 1985.

[2] C. Dodd and W. Deeds, "Analytical solutions to eddy-current probe-coil problems," *Journal of applied physics*, vol. 39, no. 6, pp. 2829–2838, 1968.

[3] K. Küpfmüller, W. Mathis, and A. Reibiger, *Theoretische Elektrotechnik*. Springer Verlag, 2006.

[4] H. Henke, *Elektromagnetische Felder*. Springer Verlag, 2001.

[5] F. Wendler, U. Tröltzsch, and O. Kanoun, "Modellierung der absoluten Impedanz einer Luftspule mit Wirbelstromrückwirkung," *Technisches Messen*, vol. 79, pp. 516–521, 2012.

[6] U. Tröltzsch, F. Wendler, and O. Kanoun, "Simplified analytical inductance model for a single turn eddy current sensor," *Sensors and Actuators A: Physical*, vol. 191, pp. 11–21, 2013.

[7] R. Ditchburn, S. Burke, and M. Posada, "Eddy-current nondestructive inspection with thin spiral coils: long cracks in steel," *Journal of nondestructive evaluation*, vol. 22, no. 2, pp. 63–77, 2003.

[8] P. Bloemen and B. Rulkens, "On the frequency dependence of the magnetic permeability of FeHfO thin films," *Journal of applied physics*, vol. 84, no. 12, pp. 6778–6781, 1998.

[9] N. Bowler, "Frequency-dependence of relative permeability in steel," in *AIP Conference Proceedings*, IOP INSTITUTE OF PHYSICS PUBLISHING LTD, vol. 820, 2006, p. 1269.

[10] K. Chowdary and S. Majetich, "Frequency-dependent magnetic permeability of $Fe_{10}Co_{90}$ nanocomposites," *Journal of Physics D: Applied Physics*, vol. 47, no. 17, p. 175–001, 2014.

[11] J. Barandiarán, A. Garcıa-Arribas, J. Muñoz, *et al.*, "Domain wall permeability limit for the giant magnetoimpedance effect," *Journal of applied physics*, vol. 91, no. 10, pp. 7451–7453, 2002.

[12] C.A. Grimes and D.M. Grimes, "Permeability and permittivity spectra of granular materials," *Phys. Rev. B*, vol. 43, pp. 10 780–10 788, 13 05/1991.

[13] C. Kittel, "Theory of the structure of ferromagnetic domains in films and small particles," *Physical Review*, vol. 70, no. 11–12, p. 965, 1946.

[14] M. Ledieu, F. Schoenstein, J.-H. Le Gallou, *et al.*, "Microwave permeability spectra of ferromagnetic thin films over a wide range of temperatures," *Journal of Applied Physics*, vol. 93, no. 10, pp. 7202–7204, 2003.

Lecture Notes on Impedance Spectroscopy, Volume 5 – Kanoun (Ed.)
© *2015 Taylor & Francis Group, London, ISBN 978-1-138-02754-1*

Impedimetric titanium dioxide thin-film gas sensors

Lucas Ebersberger, Alice Fischerauer & Gerhard Fischerauer
Chair of Metrology and Control Engineering, Universität Bayreuth, Bayreuth, Germany

ABSTRACT: Titanium dioxide, or TiO_2, is a well-known sensor material. We have investigated the impedance characteristics and surface roughness of two different types of devices involving TiO_2-coated planar interdigital electrodes. One of them has a silicon dioxide (SiO_2) insulation layer, the other one is built up without this layer. In this contribution, the response to selected hydrocarbons and humidity (at a temperature near 300 °C) and the influence of the substrate material will be discussed in some detail.

Keywords: Impedance spectroscopy; titanium dioxide; gas sensor; humidity; hydrocarbons; substrate

1 INTRODUCTION

Titanium dioxide (TiO_2) sensors are used for different applications at low and high temperatures such as oxygen sensing [1], hydrocarbon sensing [2], and NO_2 sensing [3, 4]. The material is especially interesting as it can be deposited in a reproducible manner by thin-film processes. This contribution aims at demonstrating that hydrocarbon sensors for elevated operating temperatures (about 300 °C) can be produced based on TiO_2 layers on planar interdigital electrodes made of aluminum. To this end, we will report on the sensor geometry (Section 2), the measurement setup used for sensor characterization (Section 3), and selected measurement results (Section 4).

2 SENSOR GEOMETRY

We have investigated two different types of sensor structures, called type A and B. Both consist of planar Interdigital Electrodes (IDE) made of aluminum, featuring 75 finger pairs with a finger width of 20 µm and a mark-to-space ratio of 1:1 (Fig. 1). The finger length was 4300 µm. The IDEs were fabricated by ion beam deposition of aluminum on a glass substrate. In sensor type B, the glass substrate was coated with a 100-nm SiO_2 layer before the deposition of the electrodes. This was done via reactive ion-beam sputtering of silicon.

The Al electrodes were coated with a thin (about 80 nm ± 25 nm) layer of TiO_2, also by ion beam sputtering of titanium and thermal oxidation afterwards. This oxidation was performed in an glass furnace which was heated up to 500 °C in ambient atmosphere. After two hours of monotonic heating, the temperature was held constant for an hour. After that the furnace was turned off. For a slightly faster cooling the door of the furnace was opened.

A small part on the top of the aluminum electrodes was left blank for wire bonding. The roughness, with maximum deviations from an ideally flat surface of about ±25 nm, and the layer height were measured with an Atomic Force Microscope (AFM).

The thermal postprocessing could lead to a doping of the TiO_2 layers with aluminum from the electrodes due to the absence of a diffusion barrier layer. But since similar results could be observed for reactive ion beam sputtered titanium dioxide layers (without thermal postprocessing), this influence is insignificant.

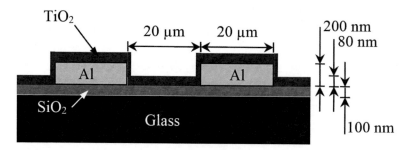

Figure 1. Schematic cross section of the sensor structure (type B). Both sensor types have 75 pairs of fingers with a length of 4300 μm.

The roughness is completely caused by the thermal treatment of the sensors. Right after the sputtering process, the roughness was negligible and increased by the thermal oxidization. This hints at a sputtering process of good quality. The increasing roughness was expected, because the oxide layers are growing from the inside (metal/silicon oxide) outward. This is a consequence of the higher diffusion rate of oxygen ions compared to titanium metal ions [5]. The increased surface could also lead to an improved sensitivity.

3 MEASUREMENT SETUP

To characterize the sensors, they were placed inside a measurement chamber, which was itself mounted in a temperature-controlled oven (Fig. 3). The atmosphere in the measurement chamber could be controlled via gas-mixing equipment involving mass-flow controllers. The chamber was first heated up to 270 °C in a nitrogen (N_2) atmosphere. When the steady state was reached, selected amounts of propane, propene and water were added to the nitrogen carrier gas. The overall gas-flow was held constant at 500 cm³/min. The analytes were pre-thinned to 800 ppm in nitrogen carrier gas. A bubbler was used for the humidification of the gas.

The sensor response to concentration steps in propane, propene, or H_2O was observed via the electrical terminals of the sensors considered as two-terminal devices. These terminal characteristics were measured by an impedance spectrometer (Impspec LF HF, Meodat) in the frequency range from 10 Hz to 5 kHz.

To discuss the results, the sensor is represented as a lossy capacitor, with both the capacitance C and the resistance R depending on frequency (Fig. 4; the frequency dependence of the equivalent-circuit elements is a consequence of the distributed nature of the processes in the sensor, which cannot be modeled appropriately by only two lumped elements with frequency-independent element values.). That simplifies the recognition of even small changes in the impedance, as changes at low frequencies become easily visible in the representation of the resistance $R(f)$ and changes at higher frequencies become even more visible in the representation of the capacitance $C(f)$.

Hence, the sensor impedance is represented as

$$\underline{Z}(f) = \frac{R(f)}{1 + j\omega R(f)C(f)} = \frac{R(f)}{1 + [\omega R(f)C(f)]^2} - j\omega \frac{R^2(f)C(f)^2}{1 + [\omega R(f)C(f)]^2}. \tag{1}$$

Some features of the impedance locus diagram depend on the gas composition in a characteristic manner. This impedance locus is described by

$$[Re\{\underline{Z}\} - 0.5 \cdot R(f)]^2 + [Im\{\underline{Z}\}]^2 = [0.5 \cdot R(f)]^2, \tag{2}$$

which would be a half-circle if R did not depend on frequency.

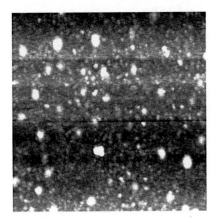

Figure 2.   AFM measurement of an area (20 μm by 20 μm) coated with titanium dioxide (sensor type A). Next to, but not onto, one of the electrode fingers.

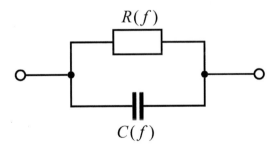

Figure 3.   Schematic drawing of the testbed used to characterize the sensors.

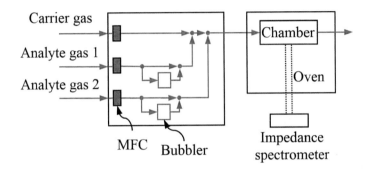

Figure 4.   Equivalent circuit for the sensors considered in this work.

When the ambient concentration of an analyte to which the sensor responds changes from c_a to $c_a + \Delta c_a$, the sensor impedance will also change. Moreover, the impedance change will lag the analyte concentration step as a consequence of molecule diffusion, charge transport etc. We may write

$$\underline{Z} = \underline{Z}(f; \Delta c_a, t). \tag{3}$$

The steady-state value of this impedance after a concentration step then is:

$$\underline{Z}_\infty(f; \Delta c_a) = \underline{Z}(f; \Delta c_a, t \to \infty). \tag{4}$$

The plots in Section 4 either visualize some feature of $\underline{Z} = \underline{Z}(f; \Delta c_a, t)$, for instance, $R(f_0; \Delta c_a, t)$ with a fixed frequency f_0 (Fig. 5), or their impedance loci $\underline{Z}_\infty(f; \Delta c_a)$ in the Nyquist plot (Fig. 6).

4   MEASUREMENT RESULTS

4.1   *Type-A sensors: Propane measurements*

Figure 5 shows the response of a type-A sensor to propane. Several features strike the eye. First, the sensor is quite sensitive: propane concentrations as low as 1 ppm are clearly detectable. However, the steady-state response saturates at the same level of $R \approx 400 \text{ k}\Omega$, irrespective of the propane concentration. This is believed to be due to the glass substrate. Such substrate influences have been identified and discussed previously for other planar sensors [6, 7]. Glass is an ionic conductor at elevated temperatures, and the equivalent resistance R_{subst} of the current paths through the glass substrate acts in parallel to the resistance $R(f)$ of the IDE sensor proper. The total resistance $R_{subst} \parallel R(f)$ can never exceed R_{subst}. This explanation was corroborated by measurements on a type-A sensor without sensitive layer.

It is also obvious that the sensor is gradually poisoned as its resistance $R_{subst} \parallel R(f)$ moves towards R_{subst} within a few days. A solution to this effect and a more proper choice for the substrate (for instance, quartz instead of glass) are prerequisites for a field use of the device. Measurements with sensors built up on quartz (not shown here) can confirm this.

The curves shown in Fig. 5 not only involve the sensor dynamics, but also the dynamics of the gas mixing apparatus, the pipes, and the measurement chamber. After the control signal to increase the propane concentration, it takes some time (less than one minute) until a concentration step (smoothed by mixing and gas flow) reaches the sensor. The associated dead times are, however, much smaller than the settling times visible in Fig. 5. Therefore, the sensor investigated exhibits the slow dynamics typical of chemical sensors in general.

Fig. 6 shows the impedance loci associated with the same measurement as the time series of Day 1 (dotted line) in Fig. 5. The curves can be classified in three groups: measurements near the steady state with a pure N_2 atmosphere, measurements near the steady state with a propane-loaded N_2 atmosphere, and measurements during the transient phase when the propane concentration in the sensor layer increases with time. The total fractional resistance change $|\Delta R(f)|/R(f)$ between the two steady states lay between -50% and 30%.

One observes that the addition of propane changes the impedance at the lower frequencies substantially. At 10 Hz, the reactance magnitude is increasing to a value about two times higher than without propane whereas, at 5 kHz, the reactance is decreasing to about two thirds of the value without propane (Fig. 6). This opposite behavior of the low- and high-frequency semicircular branches in the impedance locus diagram emphasizes the fact that the influence of the analyte gas on the sensor depends on frequency. It is not appropriate to simply state that the reactance increases or decreases with the analyte concentration. Rather, the impedance spectroscopic results suggest that one apply more detailed feature extraction methods to infer the analyte concentration and/or ambient conditions from the locus diagram.

4.2   *Type-B sensors: Humidity measurements*

With type-B sensors, even small amounts of humidity (about 1 $g(H_2O)/m^3(N_2)$) also caused a response (Fig. 7). This humidity effect on titanium dioxide sensors was to be expected and has been described in the literature [2].

The small peaks in the resistance (Fig. 7) are considered artifacts caused by switching effects of the gas-mixing equipment (especially changes of the overall gas-flow rate).

As of now, it can not be fully explained why the sensitivity to propene is close to zero in this example. It is expected, that non-stoichiometries in the oxide or different phases of the oxide could lead to that effect. Anyhow, the effect is not (completely) based on the insulation layer, because similar effects could also be observed with type-A sensors.

86

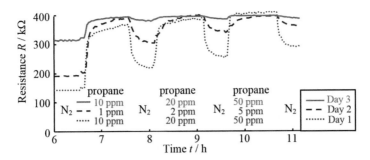

Figure 5.　Time-dependent resistance $R(f_0; \Delta c_a, t)$ at $f_0 = 10$ Hz for a type-A sensor measured three times with similar mixtures.

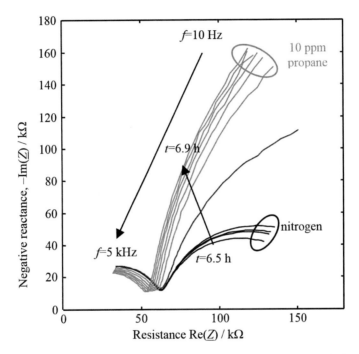

Figure 6.　Effect of a switch from pure N_2 to N_2 with 10 ppm propane (all dry) on the impedance locus diagram (measurement day 1: 6.5 h to 6.9 h, measurement period 2 minutes per curve).

The locus diagram (Fig. 8) reveals that the impedance characteristic is completely different from the dry case. The addition of the humidity lowers the apparent impedance (i.e., the modulus of the impedance) at the lower frequencies. In contrast, the high-frequency impedance is all but unchanged by the addition of humidity.

It is shown in Fig. 9 that the impedance curves of the sensor operated in a dry environment do not depend very much on time. The same can be said about the curves taken at various humidities. The results also show that the quasi-semicircular curves shrink continuously with humidity (i.e., the curves for lower humidity lie in between the curves for dry atmosphere and the curves for more humid atmosphere).

The difference of the spectral behavior observed between the measurements shown in Fig. 6 and Fig. 8, respectively, is attributed to the differences in the sensor geometry (type A vs. type B). It could be used to separate the effects of humidity and hydrocarbons. The two sensor types appear to have a pronouncedly different sensor signature in the "analyte

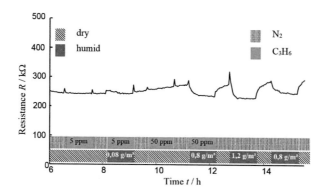

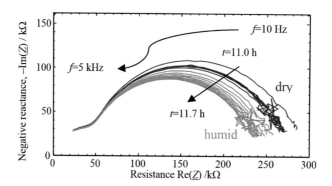

Figure 7. Time-dependent resistance $R(f_0; t; \Delta c_a, h)$ at $f_0 = 10$ Hz for a type-B sensor measured with different amounts of humidity h, expressed in g(Water)/m³(gas volume), added to pure N_2-carrier gas and C_3H_6-analyte gas.

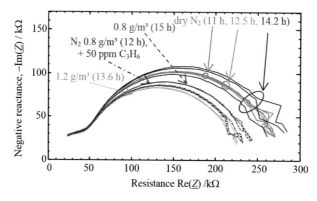

Figure 8. Effect of a switch from pure, dry N_2 to N_2 with 50 ppm propene and 0.8 g(Water)/m³(gas) on the impedance locus diagram (11.0 h to 11.7 h).

Figure 9. Details of the same measurement as in Fig. 8 at selected points. The points in time are given in the parentheses.

space" of water and hydrocarbons. This difference could be the result of the existence (sensor type B) or absence (sensor type A) of an insulating SiO_2 layer on top of the glass substrate. We attribute, however, the sensitivity to hydrocarbons to the deposited layer and the remaining impedance spectrum differences to the insulating layer.

We assume that the effect of the humidity dominates in the locus diagrams of Fig. 8 and Fig. 9, as the impedance does not change comparably when the sensor is subjected to dry propene (compare Fig. 7). Other measurements, not shown here, hint in the same direction.

5 CONCLUSION

The sensitivity of planar TiO_2-based IDE sensors to hydrocarbons and humidity was successfully demonstrated. It was shown that the response of the sensor to the analyte gas depends on frequency, which suggests the use of impedance spectroscopy to investigate the sensor mechanism at work.

It was also demonstrated that changes to the substrate, such as coating it with silicon dioxide, have a substantial impact on the impedance spectrum. Therefore, an appropriate choice of the substrate, or a substrate coating layer, helps to improve the overall sensor characteristics. For instance, the frequencies at which the impedance spectrum is most sensitive to the analyte could be shifted to values more favorable for the measurement (a shift towards higher frequencies would be tantamount to a reduction of the time needed for the measurement, which is—with the given measurement setup—about 2 min/measurement). We have shown experimentally that the sensor type investigated reacts differently to humidity (overall decrease of the apparent impedance) and to hydrocarbons (reactance magnitude increase at lower frequencies and decrease at higher frequencies). This can be exploited by feature extracting mechanisms to distinguish different analyte species.

Some issues which have not yet been resolved are the aging (poisoning) observed with the sensor elements investigated, the very details of the substrate influence effects, and the discrepancies between the responses of the type-A and type-B sensors to hydrocarbons.

ACKNOWLEDGMENT

We gratefully acknowledge support by the German Research Foundation (DFG) under contract number Fi 956/4-1.

REFERENCES

[1] U. Kirner, K. Schierbaum, W. Göpel, *et al.*, "Low and high temperature TiO_2 oxygen sensors," *Sensors and Actuators B: Chemical*, vol. 1, no. 1, pp. 103–107, 1990.

[2] D. Mardare, N. Iftimie, and D. Luca, "TiO_2 thin films as sensing gas materials," *Journal of Non-Crystalline Solids*, vol. 354, no. 35, pp. 4396–4400, 2008.

[3] A. Yüce and B. Saruhan, "Al-doped TiO_2 semiconductor gas sensor for NO_2-detection at elevated temperatures," *Tagungsband*, pp. 68–71, 2012.

[4] G. Eranna, B. Joshi, D. Runthala, et al., "Oxide materials for development of integrated gas sensors–a comprehensive review," *Critical Reviews in Solid State and Materials Sciences*, vol. 29, no. 3–4, pp. 111–188, 2004.

[5] R. Bürgel, *Handbuch Hochtemperatur-Werkstofftechnik: Grundlagen, Werkstoffbeanspruchungen, Hochtemperaturlegierungen und -beschichtungen; mit 70 Tabellen, ser.* Studium und Praxis. Vieweg, 2006.

[6] A. Fischerauer, C. Schwarzmuller, and G. Fischerauer, "Substrate influence on the characteristics of interdigital-electrode gas sensors," in *6th International Multi-Conference on Systems, Signals and Devices. SSD'09.*, IEEE, 2009, pp. 1–5.

[7] A. Fischerauer, G. Fischerauer, G. Hagen, *et al.*, "Integrated impedance based hydro-carbon gas sensors with Na-zeolite/Cr_2O_3 thin-film interfaces: From physical modeling to devices," *Physica Status Solidi*, vol. 208, no. 2, pp. 404–415, 2011.

Measurement systems

Lecture Notes on Impedance Spectroscopy, Volume 5 – Kanoun (Ed.)
© *2015 Taylor & Francis Group, London, ISBN 978-1-138-02754-1*

Miniaturized impedance analyzer using AD5933

J. Hoja & G. Lentka
Faculty of Electronics, Telecommunications and Informatics, Gdansk University of Technology, Gdansk, Poland

ABSTRACT: The paper presents the analyzer allowing to measure impedance in a range of $10\,\Omega < |Z_x| < 10\,G\,\Omega$ in a wide frequency range from 10 mHz up to 100 kHz. The device specific features are: miniaturization, low power consumptions and the use of impedance measurement method based on DSP. These features were possible thanks to the use of newest generation of large scale integration chips: e.g. "system on a chip" microsystems (AD5933) and AVR32 microcontroller. The developed analyzer metrological parameters can be compared to portable analyzers offered by worldwide manufacturers (Gamry, Ivium), but it has lower dimensions and weight, few times lower price and ability to work in distributed telemetric network.

Keywords: technical object diagnostics; impedance spectroscopy; impedance analyzer

1 INTRODUCTION

Impedance spectroscopy is used as an effective tool for testing and diagnosis of technical [1] and biological [2] objects which can be modeled by equivalent electrical circuits usually two-terminal networks [3]. Wide, continuously growing, usage of impedance spectroscopy implies growing needs for impedance analyzers, especially low-cost and field-worthy. Big companies do not offer low-cost, field-worthy impedance analyzers. Till now, leading companies: Solartron, Agilent, Novocontrol, Zahner have manufactured laboratory analyzers (e.g. FRA1260 [4], Agilent4294A [5], ALPHA [6], IM6 Workstation [7]) which were rather expensive. In last years, cheaper, portable, able to work in the laboratory as well as directly in the field, instrumentation has been proposed (Gamry Reference 600 [8], Ivium Compact Stat [9]).

When analyzing worldwide market of impedance spectroscopy instrumentation and comparing it to wide range of possible applications of impedance spectroscopy, it can be noticed that the nowadays portable impedance analyzer do not fulfill the current requirements neither functionally, nor costly nor dimensionally especially in case of test performed directly in the field including multi-point telemetric measurements. This leads to conclusion, that there is a need for miniaturized, low-cost and field-worthy impedance analyzers. Presented in the paper the impedance analyzers answer the existing needs.

2 IMPEDANCE ANALYZER ARCHITECTURE

Impedance analyzers to be able to measure different objects must be characterized by the following features:

- The impedance measurement in the grounded and non-grounded equivalent circuit,
- The possibility of free-potential compensation of electrochemical cell existing in the object under test,
- Programmed amplitude and DC offset harmonic excitation,
- Analyzer powering using USB interface or built-in battery,
- Controlling by PC via wire or wireless communication.

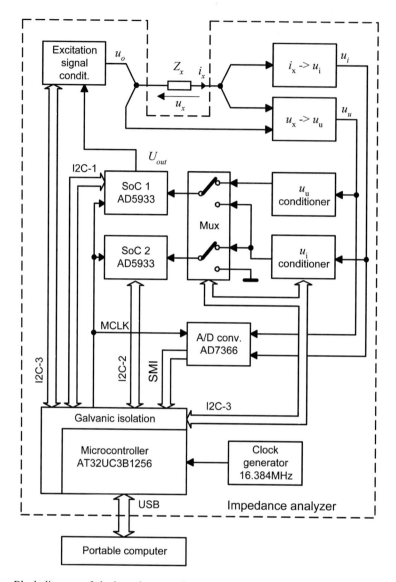

Figure 1. Block diagram of the impedance analyzer.

The block diagram of the analyzer fulfilling the above requirements is presented in Fig. 1. The realized analyzer has been featured by new solutions:

- the use of two replaceable input circuits 2 and 3 wire (2w for grounded and 3w for non-grounded impedance) for extraction of signals proportional to current through u_i and voltage u_u across the measured impedance Z_x,
- the SoC microsystems use for excitation signal generation and signals u_i and u_u orthogonal parts determination,
- he use of the newest generation microcontroller AVR32, which controls internal analyzer circuits: D/A and A/D converters, analog switches with the aid of I2C and SMI interfaces and communicates with controlling PC via USB.

Next section presents the most important solution used in analyzer–use of the SoC microsystems and microcontroller for the tested object impedance parameters determining.

3   THE CONTROLLER WIDTH SoC MICROSYSTEMS

The energy-efficient SoC microsystems AD5933 [10] have been used in analyzer for determination of orthogonal parts (Re and Im) of measurement signals u_i, u_u. AD5933 integrates sinusoidal signal generator and measurement signal orthogonal parts extraction channel using DSP technique. Calculation of parts: real (Re) and Imaginary one (Im) of the signal is performed by hardware Discrete Fourier Transform (DFT) module on the basis of signal samples acquired using 12-bit A/D converter. SoC is equipped with I2C interface used for controlling and reading internal registers (Re and Im registers).

The AD5933 manufacturer [11] does not assure possibility of driving two SoCs with common I²C-2 bus, because SoC I²C address is identical for all chips (due to this fact, in the analyzer, two separate buses I²C-1 and I²C-2 have been used, one for each SoC). Moreover, there is no full synchronous generation of excitation signal in relation to clock signal MCLK. The performed tests of the AD5933 showed that the lack of synchronization causes error of orthogonal parts (Re, Im) determination of measurement signal in SoC microsystem, which does not use own generated excitation signal [12]. This unwanted phenomena is especially visible in the range of the highest measurement frequencies 10 kHz–100 kHz.

Due to this fact, in frequency range 100 Hz–100 kHz, the signals u_i and u_u orthogonal parts are determined sequentially by SoC1 (switched by Mux). In this frequency range, measurement time is on the level of tens ms and only a little affects the impedance spectrum total acquisition time. But for frequencies below 100 Hz, simultaneous measurement of signals u_i and u_u orthogonal parts has been used with the aid of both SoC1 and SoC2. This solutions have allowed meaningful (≈ 2 times) impedance spectrum measurement time shortening.

When using its internal clock source, the AD5933 allows sinusoidal signal generation in range of 1 kHz–100 kHz. For impedance spectroscopy, the required measurement frequency range is much wider. In order to achieve impedance spectrum measurement in frequency range of 10 mHz–100 kHz, in the developed analyzer, the external clock source was used (16.384 MHz) as well as programmable frequency dividers in the microcontroller, providing external clock MLCK for SoCs.

4   IMPEDANCE ANALYZER CONSTRUCTION

The impedance analyzer was realized in module form, depending on version, on 4 or 5 PCBs packets (in version with wireless communication and battery powered) located in sealed case. The following packets: excitation signal conditioner, measurement signals conditioner, microcontroller based controller and two versions of input circuitry (depending on grounded or non-grounded objects) were designed.

The analyzer with USB communication is shown in the photo (Fig. 2). When the required distance from controlling PC is above 5 m, the alternative version of the analyzer with wireless

Figure 2.   View of internal construction of the analyzer powered and controlled via USB, showing placement of the packets in the case.

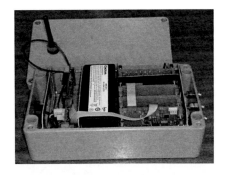

Figure 3.   View of internal construction of the analyzer controlled via wireless ZigBee power with built-in accumulator.

ZigBee based telemetry and battery-powered was designed (Fig. 3). The analyzer, besides above mentioned packets, contains additional packet responsible for ZigBee communication between the analyzer and controlling PC.

5 LABORATORY TESTS

The laboratory condition tests have been performed both on model circuit built with the reference components as well as on real anticorrosion coating samples.

At first as a test engine, a model circuit with structure and components values presented in Fig. 4. The model represents an example of equivalent circuit of anticorrosion coating in the early stage of undercoating rusting.

The measurement were performed as a comparative one using Gamry Reference 600 [8] as a reference instrument. The results of the measurement were presented in Fig. 5. Due to high impedance of the object (especially for frequencies below 1 kHz, impedance modulus goes above 1 MΩ), the measurement were performed inside the Faraday cage.

To precisely evaluate the device and using well-known (measured separately circuits components), the measurement errors were calculated (in reference to theoretical impedance value for each frequency calculated on the basis of components values) and presented in Figs. 6 and 7. Analyzing the results, it can be noted, that the obtained measurement accuracy for the presented analyzer (±2% in case of modulus and ±1.5° in case of argument) is good enough for measurement in the field and is comparable to commercially available instrumentation. The higher modulus error for Gamry instruments appears in the region were the difficult

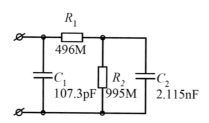

Figure 4.   Test circuit structure and components values.

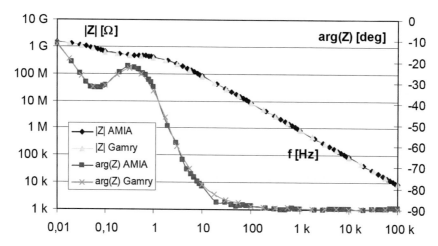

Figure 5.   Results of tests on model circuit for the presented impedance analyzer (AMIA) and the Reference 600 (Gamry).

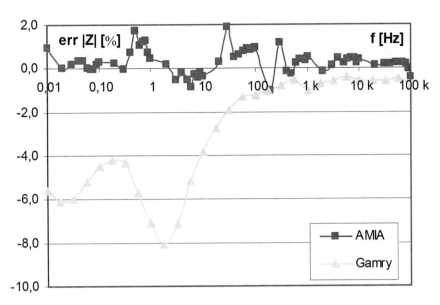

Figure 6. Relative error of impedance modulus of test circuit for the presented impedance analyzer (AMIA) and the Reference 600 (Gamry).

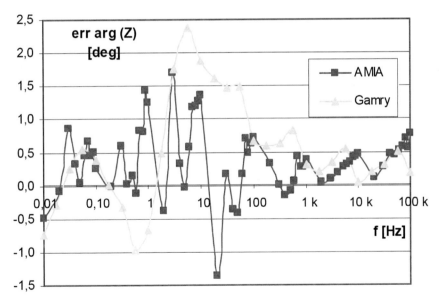

Figure 7. Absolute error of impedance argument of test circuit for the presented impedance analyzer (AMIA) and the Reference 600 (Gamry).

measurement condition exits (fast change of the impedance argument), and the phase sensitive detection is the most difficult to perform.

In the next step, the measurement on the sample of an anticorrosion coating were performed using the presented impedance analyzer and the Gamry Reference 600. The experimental setup was presented in Fig. 8. As the coating impedance is very high, the Faraday cage use was essential.

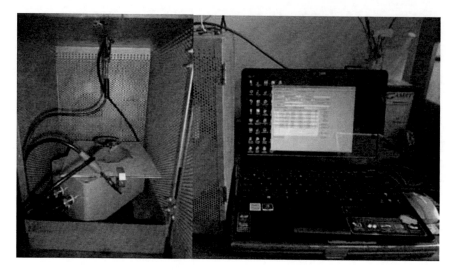

Figure 8. Laboratory setup for impedance measurement of anticorrosion coating sample using the presented analyzer (in the cages below the sample) and the Reference 600 (on the right to the PC).

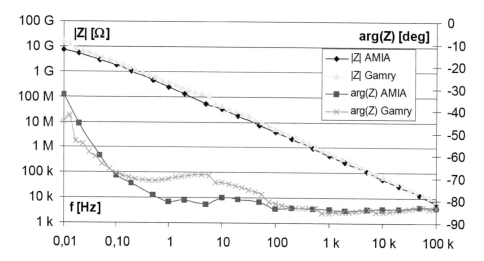

Figure 9. Impedance modulus and argument of the tested coating obtained using the presented analyzer (AMIA) and Reference 600 (Gamry).

The results of the coating measurement were presented in Fig. 9. The obtained results are very similar especially in case of impedance modulus. The differences between both instruments in case of impedance argument (in 0.2–100 Hz frequency range) require further researches.

6 TESTS IN THE FIELD

The tests in the field were performed on anticorrosion coating located on steel construction of PGE Arena Stadium in Gdansk (Fig. 10). The measurements were performed using the presented analyzer as well as the Atlas 0441 High Impedance Analyzer [13]. The measurement results are presented in Fig. 11.

The obtained results using both instruments are very similar and prove the possibility of using the presented analyzer for measurements in the field.

Figure 10.   View of localization of measurement cell and instrumentation for testing of anticorrosion coating on PGE Arena Stadium steel construction.

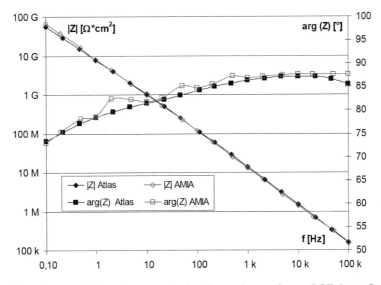

Figure 11.   Impedance modulus and argument of anticorrosion coating on PGE Arena Stadium steel construction obtained using the presented analyzer (AMIA) and the Atlas 0441 High Impedance Analyzer (Atlas).

7 CONCLUSIONS

The developed impedance spectroscopy analyzer innovative features are: miniaturization, low cost, low power consumption, possibility of controlling and powering the device via USB (or alternatively via wireless communication). The developed instrument is characterized by metrological parameters comparable to similar portable instruments offered by top worldwide manufacturers (Gamry, Ivium):

- Impedance measurement range: $10\ \Omega < |Z_x| < 10\ G\Omega$.
- Frequency range of impedance spectroscopy: from 10 mHz up to 100 kHz (10 frequency points per decade).
- Measurement signal amplitude programmable in range 1 mV–1 V (with 1 mV step), the amplitude is automatically regulated on the programmed level when changing frequency and impedance measurement range.
- DC offset set in range ±4 V with 1 mV step, allowing automatic compensation of object free-potential or extorts required DC polarization on the tested object.
- The analyzer is USB powered and consumes below 1.5 W.

On worldwide impedance spectroscopy instrumentation market, where is a lack of miniaturized, field-worthy, low-cost analyzers. Both analyzer versions are technologically advanced and can be quickly turn into medium-volume production.

ACKNOWLEDGMENTS

Authors want to thank Professor Stefan Krakowiak from Faculty of Chemistry Gdansk University of Technology for his help with measurements on coatings in laboratory.

REFERENCES

[1] P. Slepski, K. Darowicki, E. Janicka, *et al.*, "A complete impedance analysis of electrochemical cells used as energy sources," *Journal of Solid State Electrochemistry*, vol. 16, no. 11, pp. 3539–3549, 2012.

[2] Z. Xu, K.G. Neoh, and A. Kishen, "Monitoring acid-demineralization of human dentine by electrochemical impedance spectroscopy (EIS)," *Journal of dentistry*, vol. 36, no. 12, pp. 1005–1012, 2008.

[3] J.R. Macdonald and E. Barsoukov, "Impedance Spectroscopy: Theory, Experiment, and Applications," *History*, vol. 1, p. 8, 2005.

[4] Solartron. (2005). 1260 Impedance/gain-phase Analyzer. Operating Manual.

[5] Agilent. (2003). 4294A Precision Impedance Analyzer. Operation Manual.

[6] Novocontrol, *ALFA Series Analyzers*, http://www.novocontrol.de/html/index_analyzer.htm.

[7] Zahner, *Electrochemical Workstation IM6*, http://www.zahner.de/download/b_im6.pdf.

[8] Gamry Instruments, *Reference 600 Potentiostat/Galvanostat, Operator's Manual*, 2012.

[9] Ivium, *Ivium Compact Stat*, http://www.ivium.nl/CompactStat.

[10] Analog Devices. (2005). Ad5933 1 msps, 12-bit impedance converter, network analyzer.

[11] Analog Devices. (2005). Evaluation Board for the 1 MSPS 12-Bit, Impedance Converter Network Analyzer, Preliminary Technical Data EVAL-AD5933EB.

[12] J. Hoja and G. Lentka, "Portable analyzer for impedance spectroscopy," in *Proc. XIX IMEKO World Congress Fundamental and Applied Metrology, Lisbon, Portugal*, 2009, pp. 497–502.

[13] Atlas-Sollich, *ATLAS 0441 High Impedance Analyser*, http://www.atlassollich.pl/eng/products/0441.htm, 2012.

Lecture Notes on Impedance Spectroscopy, Volume 5 – Kanoun (Ed.)
© *2015 Taylor & Francis Group, London, ISBN 978-1-138-02754-1*

Electrical Cell-substrate Impedance Spectroscopy (ECIS) measurements based on Oscillation-Based Test techniques

Gloria Huertas & Andrés Maldonado
Instituto de Microelectrónica de Sevilla, CSIC, Sevilla, Spain

Alberto Yúfera
Instituto de Microelectrónica de Sevilla, CSIC, Sevilla, Spain
Department of Electronic Technology, ETSII, Universidad de Sevilla, Sevilla, Spain

Adoración Rueda & José Luis Huertas
Instituto de Microelectrónica de Sevilla, CSIC, Sevilla, Spain

ABSTRACT: A method for cell-culture real-time monitoring based on ECIS techniques, using the Oscillation-Based Test (OBT) concept is proposed. The core idea is the transformation of the Cell-Culture Under Test, CCUT into a suitable "biological" oscillator, whose characteristic parameters (frequency, amplitude, phase, etc...) are related to the cell-culture evolution and can be easily determined. An experimentally extracted cell-electrode electrical model is employed to link the oscillation parameters with the biological test to characterize the cell index in time. In summary, this work aims to provide living cell assays with a simple on-line measurement facility. Both simulation and experimental results are presented corresponding to a prototype built using discrete components and confirming the expected performances.

Keywords: Microelectrode; Bioimpedance; Cell-Culture; ECIS; Oscillation-Based Test (OBT)

1 INTRODUCTION

Electrical Cell-substrate Impedance Spectroscopy (ECIS) was described by Giaever and Keese in 1986 [1] as an interesting alternative method for experimentation in Cell Culture assays. Many works have considered the potential advantages for cell culture real-time monitoring [2], [3], allowing scientists to reduce greatly the cost and human effort in labs, which usually requires a frequent supervision as well as reproducing the assay many times to obtain intermediate samples for accounting.

The Impedance Spectroscopy (IS) applied to cell cultures measurements requires a two-electrode setup, and an AC signal source for excitation at several frequencies (normally a current), i_x at Fig. 1. From system voltage response, V_x, it is calculated the impedance (magnitude and phase) of the two-electrode plus cell culture system at a given frequency. The impedance changes are due to cell adhesion to bottom surface, where cell culture is growing, and consequently, increases with the number of cells attached to the bottom. In a two electrode setup, the measured impedance depends on the electrode technology and geometry, the electrical cell-electrode model [4]–[6] and the working frequency. Electrical measurements must consider the electrical model of the entire system to extract the useful information.

Several circuit approaches to perform impedance measurements have been proposed elsewhere [2], [3], [7], considering the need of "exciting" at a given frequency and "measuring" the response obtained. In experimental biology, it is worthwhile to develop automatic measurement techniques allowing researchers to monitor the evolution of their experiments on-line, yet requiring a simple set-up.

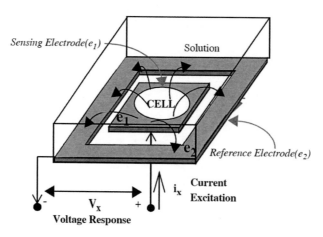

Figure 1. Two electrodes for ECIS: e_1 (sensing) and e_2 (reference). AC current i_x is injected between e_1–e_2, and voltage response V_x is measured.

This paper aims to provide living cell assays with a simple on-line measurement facility. In fact, we think that this new technical approach that facilitates monitoring of cell progression, being especially promising for real-time on-line monitoring, and thus could have an immediate impact on the biological practice. The main idea is the transformation of the *Cell-Culture Under Test*, CCUT into an oscillator, whose characteristic parameters (frequency, amplitude, phase, etc...) are related to the cell-culture evolution. By adding some extra external components, the "biological circuit" (formed by the CCUT plus the added electronic components) can be "forced" to oscillate. It is assumed that a modification in the CCUT will produce changes in either the frequency or the amplitude and consequently, any alteration will become observable. The reasons making attractive this approach are: its concept is very simple, it avoids the need of expensive equipment for stimuli generation, and the measurements to perform it are relatively simple [8]. These characteristics open the door to extend the concept of Built-In Self Test (BIST) to biological structures.

In summary, in this work, the application of OBT to biological systems is explored. An electronic (non-linear) feedback loop is used to convert the cell-electrode system into an oscillator whose oscillating parameter values (frequency, amplitude and phase) are strongly related to the cell-culture area. In this first approach, we use a simple second order band-pass filter followed by the bio-impedance and a simple comparator to establish the oscillations. The oscillating conditions are analysed using the Describing-Function (DF) method and validated through simulation with Simulink. Preliminary experimental results obtained from a discrete components prototype are also presented.

In the following, Section 2 presents the electrical model employed for the cell-electrode system characterization, and its related parameters. Section 3 introduces the OBT circuit approach and the main circuit blocks employed for testing cell culture samples. Sensitivity curves are obtained for the proposed impedance-sensing method. Simulations and experimental results illustrating the agreement of the proposed technique are described in Section 4, and finally, Section 5 summarizes the work.

2   CELL-ELECTRODE MODEL

The impedance under test of a two-electrode system in Fig. 2(a) has been already estimated [4]–[6]. In this work, it has been considered a squared gold microelectrode of 50 µm side, which can be totally or partially covered by cells in a given culture. The fill factor (ff) parameter represents the amount of the electrode's area (A) covered by cells (Ac). This is, $ff = Ac/A$.

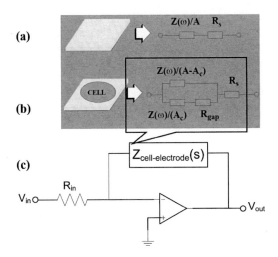

(a)

(b)

(c)

Figure 2. Bio-electronic measurement system (a) Electrical model [6], [7] of an electrode. (b) Electrical model of an electrode partially covered by cells (*ff*). (c) Circuit for cell-electrode impedance testing. Values for a $50 \times 50 \ \mu m^2$ electrode size are: $R_s = 5.4 \ k\Omega$, $Z(\omega) = C_I \| R_p$, with $C_I = 0.37 \ nF$ and $R_p = 25 \ M\Omega$. $R_{gap} = 75 \ k\Omega$.

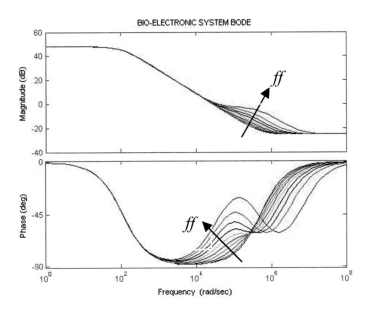

Figure 3. Bioimpedance system bode curves for several cell-culture occupation areas (*ff*).

Our objective is, using the electrical model for this system in Fig. 2(b), to obtain the cell-to-electrode area overlap (A_c) from measurements performed with the proposed circuit.

The magnitude and phase response for the $Z_{cell-electrode}(s)$ can be obtained from the set-up in Fig. 2. Resistor Rin maintains the current flowing across the cell-electrode within adequate signal levels (1–20 μA for cell protection and 10–50 mV from electrode modeling constrains [6]).

The Bode plot for the studied (in Fig. 2) bio-electronic system (magnitude and phase) is shown in Fig. 3, where curves have been considered corresponding to several cell-culture occupation areas, *ff* (this is, the microelectrode area covered by the attached cells as was above stated). Let us observe that if we consider the frequency around $(10^5/2\pi)$ Hz the system has an optimum sensitivity to *ff*. This means that can be correlated both, magnitude and phase response, with the fill factor parameter or cell-to-electrode overlap area, Ac. Most ECIS techniques search for the best frequency response for optimum impedance characterization, and then, perform the measurements knowing the *ff* dependence. Usually, absolute magnitudes (Fig. 3a) or normalized magnitudes are employed as sensitivity curves for this kind of impedance sensors.

3 OBT IMPLEMENTATION

To apply OBT to $Z_{cell-electrode}$ measurements, the cell culture under test is transformed into a robust oscillator, adding some extra components. In order to force oscillations, a positive feedback loop has to be implemented. From the point of view of OBT application, it is particularly important to accurately predict the parameters of oscillation (frequency and amplitude), analytically or by means of simulations [8]–[10]. It is also necessary to avoid the dependence of these parameters on the saturation characteristics of the active elements, like occur in common oscillators. A solution to this problem is to employ a non-linear element (a simple comparator) for the feedback loop to guarantee self-maintaining oscillations [8]–[10]. This non-linear element also allows a precise control of the oscillation amplitude. On the other hand, we need to guarantee a set of oscillation conditions in order to keep the oscillations in the nonlinear feedback loop. A simple way to implement the oscillator is using a band-pass filter in the loop, as proposed in the general circuit block proposed in Fig. 4.

In this work, the above mentioned simple scheme to implement the oscillator is considered. The non-linear feedback element is connected to the "biological filter" to implement the oscillator. In this way, only the input and output of the "biological filter" are manipulated to perform the test allowing a low intrusion in the structure.

The output of the biological filter (the input to the non-linear element) is approximately sinusoidal due to the band-pass characteristics of the global structure. This fact allows us to use the linear approximation stated in the Describing-Function (DF) method [8]–[10] for the mathematical treatment of the non-linear element.

For the sake of simplicity, let us now set up the case of a second-order bandpass filter and a comparator with saturation levels $\pm V_{ref}$. This closed-loop system verifies the required premises: the system is autonomous, the nonlinearity is both separable and frequency-independent, and the linear transfer function contains enough low-pass filtering to neglect the higher harmonics at the comparator output. Choosing adequately the band-pass filter, it can be forced that the first-order characteristic equation for the closed-loop system of Fig. 4 has an oscillation solution (ω_{osc}, a_{osc}), being ω_{osc} and a_{osc} the oscillation frequency and oscillation amplitude, respectively.

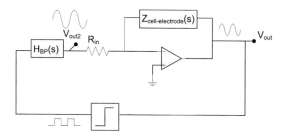

Figure 4.   Block diagram of OBT implementation measurement set-up.

Mathematically, the characteristic equation of the system of Fig. 4 would be:

$$1 + N(a)H(s) = 0 \qquad (1)$$

Being $N(a)$ the comparator Describing-Function (which can be expressed as $N(a_{osc}) = 4 \cdot V_{ref}/\pi a_{osc}$, where V_{ref} is the comparator voltage reference (whose value can be swept to obtain the adequate signal levels) and a_{osc} is the oscillation amplitude) and $H(s)$ is the transfer function of the closed-loop system (this is, the BP transfer function, HBP(s), connected in series with the SUT, $H_z(s)$, in Fig. 4). The general Band-Pass transfer function will be given by,

$$H_{BP}(s) = \frac{k_1^* \cdot \frac{\omega_0^*}{Q^*} \cdot s}{s^2 + \frac{\omega_0^*}{Q^*} \cdot s + \omega_0^{*2}} \qquad (2)$$

being ω_0^*, Q^* and k_1^* the BPF parameters. And the corresponding biological system under test transfer function (see Fig. 3), $H_z(s)$,

$$H_z(s) = \frac{k_2 \cdot s^2 + k_1 \cdot \frac{\omega_0}{Q} \cdot s + k_0 \cdot \omega_0^2}{s^2 + \frac{\omega_0}{Q} \cdot s + \omega_0^2} \qquad (3)$$

Where the constant parameters (ω_0, Q and k_0, k_1, k_2) are directly related with the electrode size, technology and biological material (ff) as given in Fig. 2 caption. Then, the global function expression will be given by,

$$H_c(s) = H_{BP}(s) \cdot H_z(s) \qquad (4)$$

To force oscillations, a pair of complex poles of the overall system has to be placed on the imaginary axis. The way to determine the oscillation conditions (gain, frequency and amplitude) is solving equation (1). This is equivalent to find the solution of this equation set.

$$1 + N(a) \cdot H(s) = (s^2 + \omega_{osc}^2) \cdot (s^2 + B \cdot s + A) = 0 \qquad (5)$$

where the coefficients are given by,

$$B = \frac{\omega_0}{Q} + \frac{\omega_0^*}{Q^*} + N(a_{osc}) \cdot k_1^* \cdot k_2 \cdot \frac{\omega_0^*}{Q^*}$$

$$A + \omega_{osc}^2 = \omega_0^2 + \frac{\omega_0}{Q} \cdot \frac{\omega_0^*}{Q^*} + \omega_0^{*2} + N(a_{osc}) \cdot k_1^* \cdot \frac{\omega_0^*}{Q^*} \cdot k_1 \cdot \frac{\omega_0}{Q} \qquad (6)$$

$$B \cdot \omega_{osc} = \frac{\omega_0^*}{Q^*} \cdot \omega_0^2 + \frac{\omega_0^*}{Q^*} \cdot \omega_0^{*2} + N(a_{osc}) \cdot k_1^* \cdot \frac{\omega_0^*}{Q^*} \cdot k_0 \cdot \omega_0^2$$

$$\omega_{osc}^2 \cdot A = \omega_0^2 \cdot \omega_0^{*2}$$

For the example, using the values of Fig. 2 (for a $50 \times 50 \ \mu m^2$ electrode size), there exists an oscillatory solution for each fill factor (ff) value. As a consequence, the main oscillation parameters are function of the occupied cell-culture area, as it is shown by theoretical predictions in Fig. 5, for the frequency and amplitude of the oscillations.

Let us observe how the oscillation frequency increases monotonically in the range [7560, 7920] Hz (0.16 Hz/μm^2 of electrode area occupied by cells) and the oscillation amplitude [0, 50] mV, when the cell-electrode area overlap (A_c) is increasing from 0 to 1. Owing that the signal level in V_{out} (see Fig. 4) is very small, as an electrode modelling constraint, it can be considered the secondary output (V_{out2} in Fig. 4) as the potential voltage output (which value is related with the oscillation frequency), thus improving the dynamic range.

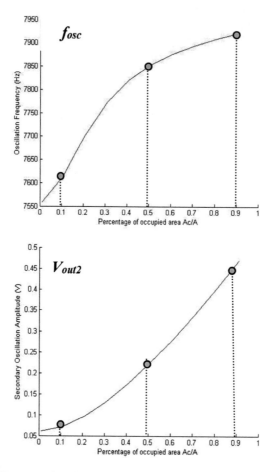

Figure 5. Theoretically expected oscillation parameters obtained from cell-to-electrode area overlap defined by $ff = A/A_c$ in the range of [0, 0.9]. The approximated sensitivities are 0.16 Hz/μm² for f_{osc}, and 0.2 mV/μm² for V_{out2}, using a squared micro-electrode of 50 μm side.

4 RESULTS

With the aim to validate the results obtained from the above-mentioned mathematical procedure, we implemented the oscillators in Simulink. A comparison between theoretical predictions and simulation results let us determine that both are very similar. It can be seen that the adopted methods predict the oscillation parameters with enough precision for the intended applications. The obtained results are summarized at Table 1 for $ff = 0.1$, 0.5 and 0.9. The obtained frequency spectrum is shown in Fig. 6 for the case $ff = 0.9$.

The complete system for ECIS test based on OBT technique was also implemented using on-board components (opamps, resistors and capacitor) [14]. Its block diagram is displayed on Fig. 7. It is composed by a Band-Pass filter, a comparator and the Z_x emulator as the main circuit blocks. Also, an Automatic Gain Control (AGC) circuit was added for limiting the voltage amplitude at the micro-electrode, and a High Pass filter for offset reduction at the comparator input voltage. The experimental Band Pass filter transfer function obtained during the test is at Fig. 8, having a center frequency of 7873 Hz. To emulate the cell culture, the values of the microelectrode model (Fig. 2a) were modified as correspond to several cases of the fill factor (0.1, 0.5 and 1.0), and then, the system was measure for each one. Voltage signals, V_{out} and V_{out2}, were measured and their voltage waveforms for $ff = 0.1$, 0.5 and 1.0, displayed at Fig. 9. From these results, the derived f_{osc} frequencies and V_{out2} amplitudes are

Table 1. Simulated oscillation parameters.

ff	f_{osc}[kHz]	V_{out} [mV]	V_{out2} [mV]
0.1	7.55	3.27	57.02
0.5	7.78	9.48	155.04
0.9	7.93	40.56	447.60

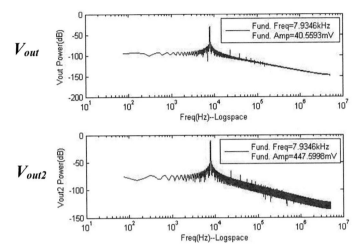

Figure 6. Simulated frequency spectrum for V_{out} and V_{out2}, using $ff = 0.9$. The oscillation parameters are $(f_{osc}, V_{out}, V_{out2}) = (7.93 \text{ kHz}, 40.56 \text{ mV}, 447.60 \text{ mV})$.

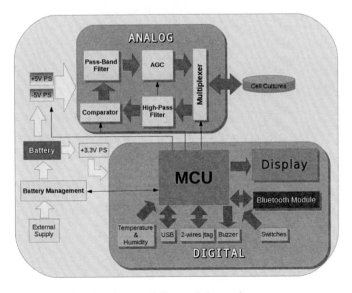

Figure 7. Block diagram of the implemented discreted electronic system.

(7600 Hz, 7705 Hz, 7940 Hz) and (83 mV, 208 mV, 430 mV) respectively. These data establish that the proposed OBT-based method for ECIS technique implementation delivers excellent sensitivities on the oscillation frequency and voltage amplitude, allowing us to evaluate the microelectrode area covered by cells in a cell culture, and to positively inform about the cell number within the culture.

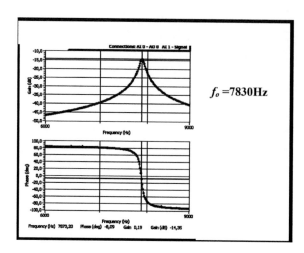

Figure 8.   Experimental BP filter transfer function.

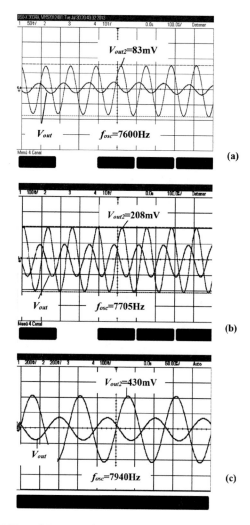

Figure 9.   Experimental V_{out} and V_{out2} waveforms for the cases: (a) $ff = 0.1$ (b) $ff = 0.5$ and (c) $ff = 1.0$. The oscillation frequency f_{osc} measure is displayed.

5 CONCLUSION

It is presented a first approach for low-cost, real-time monitoring of biological cell growth by means of an OBT methodology, based upon the use of the Describing Function (DF) technique. The proposal herein predicts with enough accuracy the frequency and amplitude of the oscillations and the required gain. Our simulation results using Simulink validate the predictions of the DF method. A first board prototype implemented with discrete components has been fabricated and tested, delivering excellent sensitivity curves for cell-culture monitoring and cell assays characterization. The procedure used here is easy to implement without the cumbersome handling of complex circuitry, and can be generalized and adapted to most impedance measurement problems.

ACKNOWLEDGEMENTS

This work was supported in part by the Spanish founded Project: TEC 2011-28302 and the Andalusian Government project: P09-TIC-5386, co-financed with FEDER program.

REFERENCES

[1] O.G. Martinsen and S. Grimnes, *Bioimpedance and Bioelectricity Basics*, 2nd ed. Elsevier Science, 2008.

[2] J. Wissenwasser, M.J. Vellekoop, and R. Heer, "Signal Generator for Wireless Impedance Monitoring of Microbiological Systems," *IEEE Transactions on Instrumentation and Measurement*, vol. 60, no. 6, pp. 2039–2046, 2011.

[3] R.D. Beach, R.W. Conlan, M.C. Godwin, *et al.*, "Towards a Miniature Implantable in Vivo Telemetry Monitoring System Dynamically Configurable as a Potentiostat or Galvanostat for Two- and Three-Electrode Biosensors," *IEEE Transactions on Instrumentation and Measurement*, vol. 54, no. 1, pp. 61–72, 2005.

[4] A. Yúfera, A. Rueda, J.M. Munoz, *et al.*, "A Tissue Impedance Measurement Chip for Myocardial Ischemia Detection," *IEEE Transactions on Circuits and Systems I: Regular Papers*, vol. 52, no. 12, pp. 2620–2628, 2005.

[5] S.M. Radke and E.C. Alocilja, "Design and Fabrication of a Microimpedance Biosensor for Bacterial Detection," *IEEE Sensors Journal*, vol. 4, no. 4, pp. 434–440, 2004.

[6] I. Giaever and C.R. Keese, "Use of Electric Fields to Monitor the Dynamical Aspect of Cell Behavior in Tissue Culture," *IEEE Transactions on Biomedical Engineering*, no. 2, pp. 242–247, 1986.

[7] D.A. Borkholder, "Cell based Biosensors Using Microelectrodes," PhD thesis, Stanford University, 1998.

[8] X. Huang, D. Nguyen, D.W. Greve, *et al.*, "Simulation of Microelectrode Impedance Changes Due to Cell Growth," *IEEE Sensors Journal*, vol. 4, no. 5, pp. 576–583, 2004.

[9] A. Manickam, A. Chevalier, M. McDermott, *et al.*, "A CMOS Electrochemical Impedance Spectroscopy (EIS) Biosensor Array," *IEEE Transactions on Biomedical Circuits and Systems*, vol. 4, no. 6, pp. 379–390, 2010.

[10] A. Yúfera and A. Rueda, "Design of a CMOS closed-loop system with applications to bio-impedance measurements," *Microelectronics Journal*, vol. 41, no. 4, pp. 231–239, 2010.

[11] P. Daza, A. Olmo, D. Cañete, *et al.*, "Monitoring Living Cell Assays with Bio-Impedance Sensors," *Sensors and Actuators B: Chemical*, vol. 176, pp. 605–610, 2013.

[12] G. Sánchez, *Oscillation-Based Test in Mixed-Signal Circuits*, ser. Frontiers in Electronic Testing. Springer, 2006.

[13] P.E. Fleischer, A. Ganesan, and K.R. Laker, "A Switched Capacitor Oscillator with Precision Amplitude Control and Guaranteed Start-Up," *IEEE Journal of Solid-State Circuits*, vol. 20, no. 2, pp. 641–647, 1985.

[14] J.A. Maldonado-Jacobi, J. Normando, A. Yúfera, *et al.*, "Cell-Culture Real-Time Monitoring Implementation," Texas Instruments, Tech. Rep., 2013.

Lecture Notes on Impedance Spectroscopy, Volume 5 – Kanoun (Ed.)
© *2015 Taylor & Francis Group, London, ISBN 978-1-138-02754-1*

High speed impedance spectroscope for metal oxide gas sensors

Marco Schüler, Tilman Sauerwald, Johannes Walter & Andreas Schütze
Laboratory for Measurement Technology, Department of Mechatronics, Saarland University,
Saarbrücken, Germany

ABSTRACT: We present a high speed, low cost impedance spectroscope for metal oxide gas sensors. It uses an MLS (Maximum Length Sequence) excitation signal, provided by an FPGA (Field Programmable Gate Array), which is also used for data acquisition. To determine the impedance spectra, we use the ETFE (Empirical Transfer Function Estimate) method, which is used to calculate the impedance by evaluating the Fourier transformations of current and voltage. Thereby, an impedance spectrum with a range from approx. 60 kHz to 100 MHz can be acquired in approx. 16 µs. For the further development of the measurement principle a simulation of the transfer function was carried out using Simulink. Using the spectrometer we built a measurement system for combined TCO (Temperature Cycled Operation)—EIS (Electrical Impedance Spectroscopy) measurements. In such a combined system, impedance spectra should be acquired in a very short time, in which the sensor can be regarded as time-invariant. The system has been tested in exemplary gas measurement applications.

Keywords: metal oxide gas sensor; high-speed impedance spectroscopy

1 INTRODUCTION

Metal oxide (MOX) gas sensors are highly sensitive to a broad range of reducing and oxidizing gases and available at relatively low cost. Standard operation and read-out are also simple, i.e. constant heating voltage for operation at elevated temperatures (typically 200 to 500°C) and measurement of the sensing layer resistance. However, low selectivity and susceptibility to poisoning (irreversible chemical reactions at the sensor's surface) and drift inhibit their use in many applications. For different measurement frequencies or temperatures, MOX gas sensor responses vary towards different gases in a characteristic manner [1]–[3]. Both Electrical Impedance Spectroscopy (EIS) and Temperature Cycled Operation (TCO) enhance selectivity and reliability of MOX gas sensors [2], [4], [5]. Selectivity is increased by the use of multidimensional data obtained at different frequencies (for EIS) or temperatures (for TCO), thus realizing a virtual multisensor. In a previous work we already described that TCO and EIS can also be used simultaneously, providing redundant system information [6].

It seems reasonable to assume that the increased selectivity is based on different processes, i.e. thermally activated surface reactions for TCO and frequency dependent relaxation of surface states for EIS. The results of these studies show that a discrimination by LDA yields different results for a poisoned sensor, depending on whether the LDA is carried out for TCO or EIS data [6]. Hence, the combined use of TCO and EIS, which is illustrated in Figure 1, can improve the reliability by comparing results, e.g. gas identifications [6], [7]. In those studies, EIS measurements had been carried out with a commercial impedance analyzer, which excites the sensor with different frequencies and records magnitude and phase of the impedance. This requires several seconds. While the results achieved with this approach are promising, further improvements are expected by its further systematic integration with TCO. In this context, measurement time is an important issue, because thermal time constants are in the order of seconds to milliseconds (for microstructured gas sensors). Thus, we favor a measurement system which can acquire an impedance spectrum in less than 1 millisecond in order to record EIS signals during temperature cycling.

The most interesting frequencies range from some 100 kHz to some 10 MHz—while the impedance is nearly constant at frequencies below 100 kHz, the lower MHz range of the impedance spectra provides features which turned out to be valuable for gas discrimination [7].

For this frequency range impedance analyzers are available, which evaluate the impedance in the frequency domain [8]. This is realized by either applying a sine voltage signal of fixed frequency and amplitude, and measuring phase and gain of the resulting current, or vice versa. Such instruments provide great benefits to laboratory-bound sensor impedance studies due to their high accuracy. On the other hand, high cost and inherently longer measurement times (e.g. more than 1 second for 400 data points in the frequency range of interest) [8] are important disadvantages for their use in a temperature-cycled gas measurement system.

By simultaneously exciting a broad range of frequencies, Fourier-based impedance analyzers have inherently shorter measurement times [9], [10], but most are optimized for the range below 1 MHz. Other issues are the relatively large dimensions of most commercially available systems as well as their relatively high cost.

In order to obtain an inexpensive measurement system which covers the frequency range of interest and offers short measurement times, we have developed a Fourier-based Impedance Spectroscope (FobIS) which uses an FPGA to generate a broad-frequency range MLS signal for excitation and to acquire the response signals for impedance calculation. Some of the most important requirements of this system are summarized in Table 1.

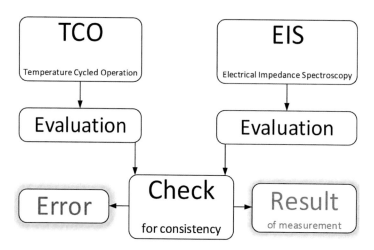

Figure 1.   Self-monitoring strategy for semiconductor gas sensors: Data from both TCO and EIS are evaluated separately and the output is checked for consistency [6].

Table 1.   Requirements for an impedance spectroscope for combined TCO-EIS operation of MOX gas sensors.

Property	Desired value	Optimisation goal	Importance
Measurement time	<1 ms	Shortest	+++
Impedance range	100 Ω ... 1 MΩ	Broadest	++
Frequency range	100 kHz ... 100 MHz	Broadest	+
Hardware cost	1000 €	Lowest	+
Phys. dimensions	Portable	Smallest	+

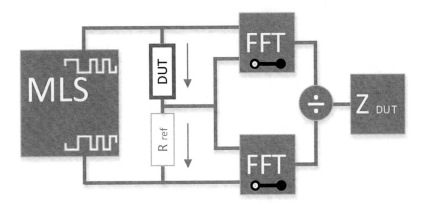

Figure 2.   Overall measurement concept for EIS based on binary signal stimulation and Fast Fourier Transform (FFT) transformation of the time domain response.

2   FOURIER-TRANSFORM BASED MEASUREMENT CONCEPT

The fundamental concept of our FobIS system was presented in [11]. It is based on the use of Empirical Transfer Function Estimate (Equation 1 [12]), which estimates a system's transfer function by dividingoutput and input in the frequency domain:

$$\widehat{G}_s(i\omega) = \frac{Y_s(\omega)}{U_s(\omega)}\tag{1}$$

The impedance of a gas sensor can be regarded as the transfer function describing the relationship between voltage and current; it can thus be calculated from the ratio of these variables in the frequency domain (Equation 2: Calculation of sensor impedance using ETFE).

$$Z(\omega) = \frac{U(\omega)}{I(\omega)} = \frac{U_{sensor}(\omega)}{\frac{U_{ref}(\omega)}{R_{ref}}}\tag{2}$$

In our system, the current is calculated from the voltage drop over a reference resistor of known value. The measurement concept is shown in Figure 2: On the left hand side, the generation of an MLS (Maximum Length Sequence) signal is indicated. This signal has high energy content in a broad range of frequencies, which is a premise for a good SNR (see "Excitation signal"). The MLS signal is applied to a voltage divider consisting of the Device Under Test (DUT), i.e. our MOX gas sensor, and a reference resistor of known impedance, which is used to determine the current (cf. Equation 2).

For calibration purposes, resistors of known impedance can be used in place of the DUT. Both voltage signals are Fourier transformed to enable calculation of the impedance according to Equation 2.

3   HARDWARE

In order to enable measurements in the range up to 100 MHz, we use a high speed ADC, which is evaluated using an FPGA. This FPGA is also used to generate a binary excitation signal via a digital output. To minimize complexity, we decided to use binary MLS (Maximum Length Sequence) signals. The choice of the excitation signal is explained in more detail in the section "Excitation signal".

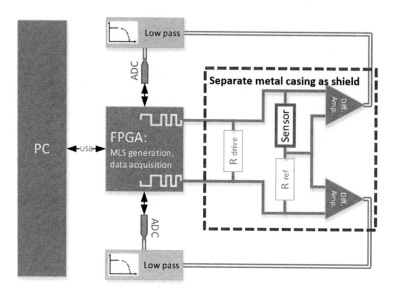

Figure 3.   Hardware setup of the FobIS low cost impedance spectroscope.

The central component of the spectroscope is an FPGA (Xilinx®Virtex-4). It generates an MLS signal through a linear shift register, implemented in Verilog. The generation of MLS signals by linear shift registers is described in detail in [13]. Alternatively, the sequence could be generated once and stored on the FPGA. In our case, this would demand $2^{11} - 1 = 2047$ bits of memory for an MLS 11 signal, whereas the generation of the sequence on the FPGA only requires a 3-step linear shift register with a width of 11 bit and a corresponding XOR gate [13]. Thus, the shift register was realized on the FPGA to generate the sequence in real time. The hardware setup of the system is depicted in Figure 3. As indicated here, the signal voltage is generated by the voltage drop over a resistor (R_{drive}) with a value of 75 Ω. The MLS signal is transmitted to this resistor using the BLVDS_25 differential signaling standard, which provides a current of defined value (16 mA). This setup results in an amplitude of 1.2 V, which was chosen to provide a satisfying SNR and stay within the ADC input range (2 V). The amplitude can be varied by either adjusting the signal current provided by the FPGA or by replacing R_{drive}.

The signal is applied to a voltage divider which is connected in parallel to R_{drive}. This voltage divider consists of the sensor and a high-precision 10 kΩ reference resistor. Both branches of the voltage divider are connected to TI (Texas Instruments) ADS62P49EVM high-speed ADCs. In order to prevent distortions by the relatively low input impedances of the ADC, the signals are stabilized by differential amplifiers (AD 8130) with an input impedance of 1 MΩ, which are placed in direct vicinity to the measured branches. The amplifiers and the voltage divider are encased in a separate metal housing to suppress electromagnetic interferences. Besides the choice of a particular binary sequence, the frequency content of the signal is strongly influenced by the hardware setup. It should have high energy content in the frequency range which is of interest for measurement, but low energy content above the system's Nyquist Frequency to minimize frequency aliasing. 7th order Butterworth low pass filters with a cutoff frequency of 100 MHz are therefore placed before the ADCs to eliminate frequencies above the range of interest.

4   EXCITATION SIGNAL

Within this work, we are using MLS (Maximum Length Sequence) signals, also called MLBS (Maximum Length Binary Sequences), m-sequences, or maximum-length linear shift register sequences, which can be easily generated using a simple shift register [13], [14]. The generation of an MLS signal of length $2^n - 1$ ($n \geq 3 \in \mathbb{N}$) can be realized with a linear shift register of n bit length [13]. This allows very efficient implementation in an FPGA. MLS have some properties that make them a suitable choice for ETFE-based measurements. The signal energy input of a voltage drop U across an impedance Z is as follows:

$$E_{Signal} = \int_0^{T_{meas}} \frac{U(t)^2}{Z} dt = \frac{1}{Z} \int_0^{T_{meas}} U(t)^2 dt \tag{3}$$

The crest factor C is a measure for the ratio between peak amplitude and Root Mean Square (RMS) value of a signal:

$$C = \frac{|U_{Peak}|}{U_{RMS}} \tag{4}$$

It is therefore inversely proportional to the signal's energy-to-amplitude ratio, which means that a signal with low crest factor contains a high signal power for a given duration and amplitude. For a given noise level, the energy content of an excitation signal proportionally determines the Signal to Noise Ratio (SNR) of the system. While sine signals, including sine sweeps/chirps have a satisfying crest factor of $\sqrt{2}$, the crest factor is ideal, i.e. 1, for symmetrical square wave signals, e.g. rectangular chirps and the MLS signal.

Another important reason for the choice of this signal is the fact, that its energy seems to be distributed uniformly over the frequency spectrum: the absolute value of its discrete Fourier transform (DFT) is perfectly uniform as shown in Figure 4(a).

Since we use FFT in our FobIS system for computational efficiency, and thus need a signal of length 2^n ($n \in \mathbb{N}$), we have to pad the signal with an additional zero (or negative voltage for the differential signal). The energy content of the zero-padded signal is shown in Figure 4(b). Evidently, this signal does not have a uniform energy at all. This is due to the fact that the energy content of an MLS signal is uniform at the discrete sampling frequencies from f_S/L to $(1 - \frac{1}{L}) \cdot f_S/2$ with a uniform frequency separation of f_S/L, i.e. the harmonics of the MLS signal, but not between those frequencies [15], [16]. This is demonstrated in Figure 5, which compares a section of the DFT of an MLS of length 511 to the corresponding DFT of the same signal, which was zero-padded to 10-fold length in order to achieve a 10-fold frequency resolution. The fluctuations in the highly resolved DFT show that the energy content is not uniform at frequencies which are not harmonics of the unpadded MLS.

Using a signal with such large energy fluctuations will cause strong variations in SNR amongst frequencies; even worse, it can be assumed that nonlinearities will have a strong effect due to the steepness of the fluctuations. In order to find a signal providing good measurement results, we simulated the system. Figure 6 illustrates our approach to this simulation.

We generate a signal in the time domain, which is then fed into the Transfer Functions (TF) modelling the relationship between the voltage input and the voltage across the sensor on the one hand and the voltage input and the voltage across the reference resistor on the other. This calculation is performed in the time domain using the Bogacki-Shampine algorithm in Mathworks Simulink. The transfer functions are based on an empirically determined equivalent circuit of the sensor, which is shown in Figure 7. This equivalent circuit has been shown to be an appropriate model of the broad range gas sensor UST 1330 (UST Umweltsensortechnik GmbH) [11]. The transfer functions derived from this circuit are as follows:

$$TF_{Sensor} = \frac{U_{sensor}}{U_{in}} = \frac{R \cdot L \cdot C \cdot s^2 + L \cdot s + R}{R \cdot R_{ref} \cdot L \cdot C \cdot C_p \cdot s^3 + (R \cdot L \cdot C + R_{ref} \cdot L(C + C_p)) \cdot s^2 + (L + R \cdot R_{ref} \cdot C_p) \cdot s + R_{ref} + R} \tag{5}$$

$$TF_{R_{ref}} = \frac{U_{R_{ref}}}{U_{in}} = \frac{R \cdot R_{ref} \cdot L \cdot C \cdot C_p \cdot s^3 + R_{ref} \cdot L(C + C_p) \cdot s^2 + R \cdot R_{ref} \cdot C_p \cdot s + R_{ref}}{R \cdot R_{ref} \cdot L \cdot C \cdot C_p \cdot s^3 + (R \cdot L \cdot C + R_{ref} \cdot L(C + C_p)) \cdot s^2 + (L + R \cdot R_{ref} \cdot C_p) \cdot s + R_{ref} + R} \tag{6}$$

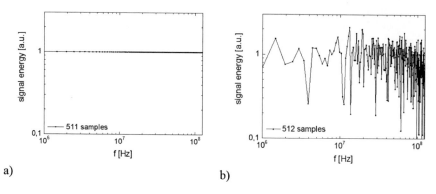

a)   b)

Figure 4.   Absolute values of the DFT for an MLS signal of length 511, calculated for 511 (a) and 512 (b) samples.

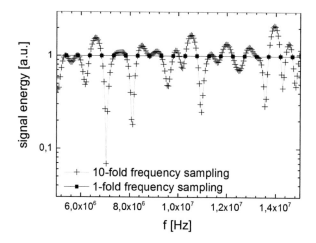

Figure 5.   Energy content of an MLS signal in and between its harmonics.

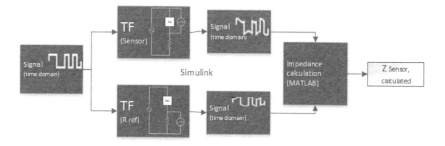

Figure 6.   Schematic of simulation approach.

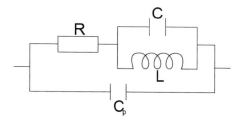

Figure 7.   Equivalent circuit of the MOX gas sensor determined empirically [11].

By time-domain based application of these two transfer functions, the output voltage signals are calculated. These are then fed to the impedance calculation algorithm, which calculates the impedance based on the method which is used for the measurements. With this method it is possible to numerically determine the influence of the excitation signal in the actual measurement algorithm.

Figure 8 shows the transfer function (a) as well as simulation results for three cases (b–d). In the first case we use an excitation signal of $2^{12} - 1$ points (MLS12, generated from the primitive polynomial $1 + x + x^4 + x^6 + x^{12}$) and a DFT of $2^{12} - 1$ points. The generation of MLS signals from given primitive polynomials is described in [11], [17]. The simulated transfer function which was calculated using this excitation signal (Figure 8b) is very smooth and closely resembles the true transfer function. In the second case the same excitation signal is used, but padded with one additional zero to obtain 2^{12} samples in order to perform an FFT. The simulated transfer function (Figure 8c) has roughly the same shape as the true transfer function but contains strong distortions, which resemble noise but are actually deterministic due to the shifted sampling points of the FFT. In the third case an excitation signal of $2^{11} - 1$ points (MLS11, generated from the primitive polynomial $1 + x^9 + x^{11}$) is used, which is padded with $2^{11} + 1$ zeros to obtain an overall signal length of 2^{12} points. The resulting simulated transfer function (Figure 8d) is smooth and corresponds very well to the true transfer function. The first and second results are not particularly surprising, since the DFT of the MLS12 signal's harmonic is smooth and that of the zero-padded signal is fluctuating strongly. However, it is surprising that the fluctuations are cancelled out when the strongly padded MLS11 signal is used. Although the mechanism for the improved properties of the MLS11 signal is not fully understood, it makes sense to use the padded MLS11 signal as a suitable excitation signalfor our FobIS impedance spectroscope which can be as well generated and processed very efficiently.

The total duration of this excitation signal is 16.384 μs (212.4 ns) using a clock frequency of 250 MHz, resulting in a signal bandwidth of 61 kHz to 125 MHz. The lower boundary of this bandwidth results from the signal length and equals the signal length's inverse value:

$$f_{min} = \frac{1}{T_{signal}} = \frac{1}{16.384\ \mu s} = 61.0444\ kHz \qquad (7)$$

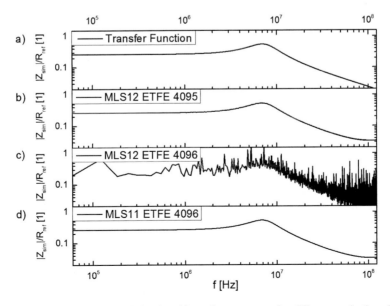

Figure 8. Transfer function (a) and simulated impedance spectra for different excitation signals and measurement lengths (b–d, see text for details).

Table 2. Frequency ranges considered within this work.

	Lower boundary	Upper boundary
Frequency range of excitation signal	≈ 61 kHz (cf. Equation 7)	≈ 125 MHz (cf. Equation 8)
3 dB cutoff frequency of anti-aliasing filter	–	≈ 100 MHz
Frequency range for gas sensor measurements (range from which features were extracted for LDA)	100 kHz … 100 MHz	≈ 100 MHz

The upper boundary results from the measurement system's Nyquist Frequency:

$$f_{max} = \frac{f_{sampling}}{2} = \frac{250 \text{ MHz}}{2} = 125 \text{ MHz} \tag{8}$$

As described in the hardware section, the measurement system does not use the entire bandwidth of this signal. Table 2 shows the spectral boundaries of the setup.

5 EXPERIMENTAL RESULTS

Our FobIS system was tested for the discrimination of different test gases using a commercial SnO_2-based broad range gas sensor [18]. Both impedance data and data obtained from temperature cycling were evaluated. To apply the self-test strategy described in the introduction, the discrimination of different gases must be possible using either type of data exclusively. The measurements were carried out in a gas mixing apparatus providing a constant air flow of 200 ml/min with constant relative humidity of 50%. The test gases measured here were methane (CH_4, 550 ppm/1100 ppm), carbon monoxide (CO, 50 ppm/100 ppm), hydrogen (H_2, 5 ppm/10 ppm) and ethanol (C_2H_5OH, 5 ppm/10 ppm). The sensor was exposed to each gas concentration admixed to pure air for a period of 20 min, followed by 20 min in pure air. The measurement was carried out during temperature-cycled operation: the temperature cycle consists of six temperature steps, with temperature set points equidistantly spread between 200°C and 450°C. The duration of the first temperature step (200°C) is 30 s, the other steps have a duration of 18 s each.

Figure 9 shows the reactions towards some reducing test gases in the ppm range, during a laboratory measurement. The data shown here are the sensitivities, i.e. $Z_{air}/Z_{gas} - 1$, at 61 kHz and 12 MHz, respectively.

Ten impedance measurements were averaged for each data point, providing a good SNR for the 61 kHz impedance data. For the 12 MHz data, the SNR is obviously lower. Due to the parasitic capacitances in the system, the SNR decreases strongly for higher frequencies. However, at least for lower sensor temperatures where the sensor has a higher sensitivity, the gas profile is still recognizable in the data acquired at 12 MHz.

Figure 10 shows impedance spectra acquired for the UST 1330 sensor at 250°C in pure air with 50% r.h. (top) compared to the sensitivity spectra acquired in 550 ppm CH_4 (middle) and 50 ppm CO (bottom), respectively. For both gases, the highest sensitivity is found at the lowest frequency (61 kHz). At the second lowest frequency (122 kHz), the sensitivity to methane is almost zero and thus differs strongly from the sensitivity to CO, which has a value of 0.64. At 244 kHz, there is a peak of the sensitivity for both gases. The sensitivity drops strongly at higher frequencies but, as shown in Figure 9, it is possible to see changes in the impedance even at 12 MHz, which can be used to compute features for gas discrimination using multivariate statistics, e.g. Principal Component Analysis (PCA) or Linear Discriminant Analysis (LDA) [19]. Averaging impedance values of adjacent frequencies can help to improve the SNR especially in the higher frequency range. The mean values acquired in this way can then be used as features for data evaluation.

118

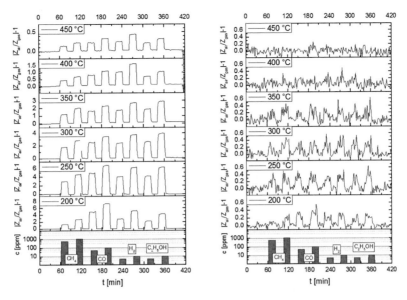

Figure 9.   Sensitivities at 61 kHz (left) and 12 MHz (right) during test gas profile.

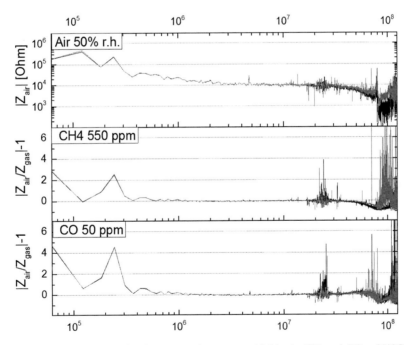

Figure 10.   Impedance spectrum in air compared to gas sensitivities in CH_4 and CO at 250°C.

The absolute accuracy of our FobIS measurement system is not comparable to commercial lab equipment, and there are important variations in SNR, depending on the impedance values and frequency measured. Nonetheless, we can use it to acquire impedance data in the field with good reproducibility. These data can be used to compute features for classification, i.e. gas discrimination. As usual in multivariate classification, features are derived from the measured raw data, e.g. mean values or slopes of the impedance spectrum or the resistance response curve in TCO [20].

Figure 11 shows the comparison of LDA projections based on features derived from TCO data (a) and EIS data (b) obtained from the measurement described above.

Fig. 11(a) is based on data acquired at 61 kHz and four different temperature levels; Figure 11(b) is based on EIS data acquired at 250°C in four different frequency ranges. Both LDAs use only four features to prevent overfitting, i.e. the poor generalization of the LDA when using a relatively large number of features [21]. The graphs show the data acquired in pure air, 550 ppm methane (CH_4), and 50 ppm carbon monoxide (CO), respectively, with the conditions described previously. The data points of the different test gases are grouped very close to each other and far apart from the other gases, thus enabling a good discrimination of the gases using either TCO or EIS data. These results show that the data acquired with the FobIS system can be used to discriminate different gases by evaluation of either temperature cycle or impedance data thus allowing the application of the sensor self-test strategy. To evaluate the self-test-strategy illustrated in Figure 1, we are currently analyzing data from several test measurements acquired with new and damaged sensors using both a commercial impedance analyzer and the FobIS system. In addition to the laboratory tests, we are carrying out a long-term field test in an underground parking garage. The measurement system used in this test has a slightly different setup: the wires connecting the voltage divider branches with the differential amplifier (see Figure 3) are connected by shielded cables, which have higher parasitic capacities.

Figure 12 shows a measurement series acquired during the time period from Sunday, Oct. 27, 2013, 13:00 through Wednesday, Oct.30, 24:00. The plots show sensitivity values at different frequencies; all graphs are based on data from the 250°C temperature step, at which the sensor is highly sensitive to carbon monoxide. In the top row, a reference measurement acquired with an electrochemical cell (Dräger Polytron, CO-Sensor 6809605) is shown. The measurement was carried out in the underground parking garage of a public building, which is mainly used by the employees. The highest CO concentrations are observed in the early evening hours, when a relatively high number of cars are started and driven out by employees leaving their offices. It should be pointed out that the reaction of the semiconductor gas sensor is relatively high compared to the CO value measured by the electrochemical reference sensor. This can be explained by the fact that the MOX gas sensor is sensitive to CO, but also to a variety of other gases, which are emitted by the vehicles together with carbon monoxide, especially unburned fuel components (hydrocarbons, HC). The absence of peaks in the morning hours when employees are arriving is due to the fact that practically all vehicles are equipped with catalytic converters resulting in very low CO and HC emissions once the vehicles are warmed up.

The measured sensor reactions correspond clearly to the values measured by the reference sensor shown at the top, which demonstrates the usability of the sensor system for field

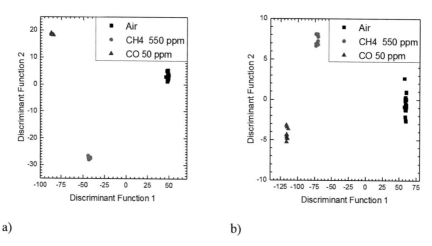

a)   b)

Figure 11.   LDA projections obtained from temperature cycle data (a) and impedance data (b) for the classification of CH_4 and CO vs. pure air.

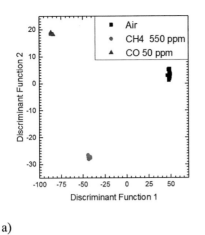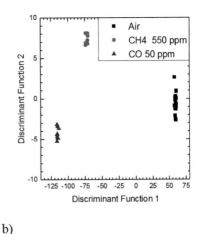

a)   b)

Figure 12. Measurement obtained over a period of three days in an underground parking garage (Sunday afternoon through Wednesday night).

measurements of, for example, *CO* and *HC*. Multivariate data analysis could be used to determine the *CO* to *HC* composition and also other gas sources, e.g. fuel leaks or fires.

6 DISCUSSION AND OUTLOOK

The impedance spectroscope developed in this work can be used to acquire impedance values in a frequency range from 61 kHz up to ca. 100 MHz. For determination of absolute impedance values, the system has to be calibrated using calibration impedances, i.e. short circuit, open end and a known resistor. In this work, we have evaluated relative values of the impedance sensor signal, which proved sufficiently reproducible for gas discrimination. A simulation of the transfer function has been implemented to optimize the excitation signal. We could show that an MLS11 signal padded with zeros provides better results than the MLS12 signal for the presented setup when using FFT to determine the impedance spectrum with low computational effort. Although the simulation does not model all aspects of the system, it is an important tool for selection of a suitable excitation signal. The consideration of further aspects, e.g. signal propagation delays, could improve its benefits and enable the choice of an even more suitable signal. To this aim, the range of excitation signals could also be expanded. Many excitation signals with different spectral properties are available [22], [23]. This could be used to improve the excitation signal.

The frequency bandwidth of the system is 61 kHz to 100 MHz, however, due to parasitic capacitances, the signal intensity decreases at high frequencies and therefore the SNR is decreasing inherently. Nonetheless, as shown in the section "Experimental results", accurate measurements of gas sensor features are possible at e.g. 12 MHz and 18 MHz. Averaging over several frequencies, the extraction of impedance features is possible at frequencies up to approx. 100 MHz. The time required for the acquisition of a single measurement is only 16.384 µs. Thus, the sensor can be regarded as a time-invariant system during the data acquisition and even averaging over several measurements is possible to improve the signal quality. Currently the data transmission to a PC is the most time consuming step limiting the sampling rate to one spectrum per second. To avoid this we plan to perform the full data processing within the FPGA. With dimensions of 370×270 times 110 mm^3, it is portable and can thus easily be used for field application. The hardware cost is approx. 1000 €, and lies within the requirements. By further integration, size, cost and power consumption can be further reduced in order to meet the requirement for integration into a sensor system.

ACKNOWLEDGEMENTS

We gratefully acknowledge funding by the German ministry of Economic Affairs and Energy (BMWi, grant no. 16962N). The authors would like to thank Drägerwerk AG & Co. KGaA, Lübeck, Germany and 3S GmbH, Saarbrücken, Germany, for hardware support.

REFERENCES

[1] T. Sauerwald, D. Skiera, and C.-D. Kohl, "Selectivity enhancement of gas sensors using non-equilibrium polarisation effects in metal oxide films," *Applied Physics A*, vol. 87, no. 3, pp. 525–529, 2007.

[2] U. Weimar and W. Göpel, "AC measurements on tin oxide sensors to improve selectivities and sensitivities," *Sensors and Actuators B: Chemical*, vol. 26, no. 1, pp. 13–18, 1995.

[3] S.R. Morrison, "Semiconductor Gas Sensors," *Sensors and Actuators*, vol. 2, pp. 329–341, 1982.

[4] A. Schütze, A. Gramm, and T. Rühl, "Identification of Organic Solvents by a Virtual Multisensor System with Hierarchical Classification," *Sensors Journal, IEEE*, vol. 4, no. 6, pp. 857–863, 2004.

[5] P. Reimann and A. Schütze, "Fire detection in coal mines based on semiconductor gas sensors," *Sensor Review*, vol. 32, no. 1, pp. 47–58, 2012.

[6] P. Reimann, A. Dausend, and A. Schutze, "A self-monitoring and self-diagnosis strategy for semi-conductor gas sensor systems," in *Sensors, 2008 IEEE*, IEEE, 2008, pp. 192–195.

[7] T. Conrad, F. Trümper, H. Hettrich, *et al.*, "Improving the Performance of Gas Sensor Systems by Impedance Spectroscopy: Application in Under-Ground Early Fire Detection," in *Proceedings of the SENSOR Conference*, AMA Service GmbH, vol. 1, 2007, pp. 169–74.

[8] Agilent, *Agilent 4294A Precision Impedance Analizer Data Sheet*, http://cp.literature.agilent.com/litweb/pdf/5968-3809E.pdf, retrieved 2013/11/25.

[9] Solartron, http://www.solartronanalytical.com/download/ModuLab-MTS.pdf, retrieved 2013/11/25.

[10] Eliko, http://www.eliko.ee/impedance-applications/, retrieved 2013/11/ 25.

[11] P. Reimann, M. Schüler, S. Darsch, *et al.*, "Hardware concept and feature extraction for low-cost impedance spectroscopy for semiconductor gas sensors," *Lecture Notes on Impedance Spectroscopy*, vol. 2, pp. 905–907, 2012.

[12] L. Ljung and T. Glad, "Modeling of dynamic systems," 1994.

[13] J. Borish and J.B. Angell, "An Efficient Algorithm for Measuring the Impulse Response Using Pseu-dorandom Noise," *Journal of the Audio Engineering Society*, vol. 31, no. 7/8, pp. 478–488, 1983.

[14] S.W. Golomb, "Shift-Register Sequences and Spread-Spectrum Communications," in *Third International Symposium on Spread Spectrum Techniques and Applications, 1994. IEEE ISSSTA'94*, IEEE, 1994, pp. 14–15.

[15] J. Vanderkooy, "Aspects of MLS Measuring Systems," *Journal of the Audio Engineering Society*, vol. 42, no. 4, pp. 219–231, 1994.

[16] R. Mazurek and H. Lasota, "Application of Maximum-Length Sequences to Impulse Response Meas-urement of Hydroacoustic Communications Systems," *Hydroacoustics*, vol. 10, pp. 123–130, 2007.

[17] N. Xiang and K. Genuit, "Characteristic Maximum-Length Sequences for the Interleaved Sampling Method," *Acustica*, vol. 82, pp. 905–907, 1996.

[18] Umweltsensortechnik, http://www.umweltsensortechnik.de, retrieved 2013/11/25.

[19] K. Backhaus, B. Erichson, W. Plinke, *et al.*, *Multivariate Analysemethoden*, Springer-Verlag, Ed. 2000.

[20] Z. Ankara, T. Kammerer, A. Gramm, *et al.*, "Low power virtual sensor array based on a microma-chined gas sensor for fast discrimination between $H2$, CO and relative humidity," *Sensors and Actuators B: Chemical*, vol. 100, no. 1, pp. 240–245, 2004.

[21] D. Luo, C.H. Ding, and H. Huang, "Linear Discriminant Analysis: New Formulations and Overfit Analysis," in *AAAI*, 2011.

[22] K. Godfrey, A. Tan, H. Barker, *et al.*, "A survey of readily accessible perturbation signals for system identification in the frequency domain," *Control Engineering Practice*, vol. 13, no. 11, pp. 1391–1402, 2005.

[23] A.H. Tan and K.R. Godfrey, "A Guide to the Design and Selection of Perturbation Signals," *in Proceedings of the 48th IEEE Conference on Decision and Control, 2009 held jointly with the 2009 28th Chinese Control Conference (CDC/CCC)*, IEEE, 2009, pp. 464–469.

[24] A. Devices, *Datasheet AD 8129/AD 8130*, http://www.analog.com/static/imported-files/data_sheets/AD8129_8130.pdf, retrieved 2013/11/25.

Lecture Notes on Impedance Spectroscopy, Volume 5 – Kanoun (Ed.)
© *2015 Taylor & Francis Group, London, ISBN 978-1-138-02754-1*

Can Impedance Spectroscopy serve in an embedded multi-sensor system to improve driver drowsiness detection?

Li Li, Thomas Bölke & Andreas König
*Department of Electrical and Computer Engineering, Institute of Integrated Sensor Systems,
TU Kaiserslautern, Germany*

ABSTRACT: Driver status monitoring belongs to the key components of active safety systems which are capable of improving car and road safety without compromising driving experience. To improve effectiveness and robustness by adding sensing capability to a prototype of driver assistance system, Impedance Spectroscopy (IS) is investigated with respect to the correlation between impedance response and driver drowsiness/vigilance. For this key aim and potential vehicle integration (via on-board equipment), suitable embedded IS-system is needed and the first-cut implementation serves here for study. The experimental results demonstrate that with IS the classification accuracy has been improved by 10% for three drowsiness levels based on data sets of five test subjects with 300-minute driving sequence.

Keywords: Impedance Spectroscopy; Multi-Sensor System; Drowsiness Detection

1 INTRODUCTION

Drowsy driving is a serious problem that can affect anyone on the road. As conservatively estimated by National Highway Traffic Safety Administration (NHTSA), 1,550 deaths, 71,000 injuries, and $12.5 billion monetary losses are the result of 100,000 police-reported crashes which are directly caused by drowsy driving each year in the United States [1]. A study by the Federal Highway Research Institute (BASt) in Germany presented that drowsy driving was the second most frequent cause of serious truck accidents on German highways [2]. Due to severe damage caused by drowsy truck or bus drivers it is urgent to extend active safety to cope with driver drowsiness in commercial vehicles. On the other hand, the automotive segment, including utility vehicles, continuously demands cheap and low-power system solution for high volume application in this field.

Driver drowsiness/vigilance can be monitored by utilizing biosignals such as ECG, EEG, EOG, eyelid movements and skin impedance [3]. Intrusive measurement techniques like EEG or EOG, which can stress or distract drivers and impact driving performance significantly, are therefore not suitable for on-board equipment in vehicle. In previous studies [4], [5] non-intrusive measurement systems embedded in steering wheel or in car seat for recording ECG signals have been demonstrated. As investigated in related work [6] the skin resistance response reflects the change of human vigilance level. In this work, to investigate the potential of IS to complement information for improved drowsiness detection we present a first prototype of embedded IS-system being integrated in the existing intelligent multi-sensor system for driver status monitoring.

After introduction a prototype of intelligent multi-sensor system for driver status monitoring—DeCaDrive is presented in Section 2. The system expansion with embedded impedance spectroscopy sensor, its analog front-end and sensor data preprocessing are addressed in Section 3. Multi-sensor feature computation and data fusion as well as neural network based pattern classification are discussed in Section 4. The extended system is validated and evaluated by presenting the experimental results in Section 5. Finally, with future perspectives the current work is concluded in Section 6.

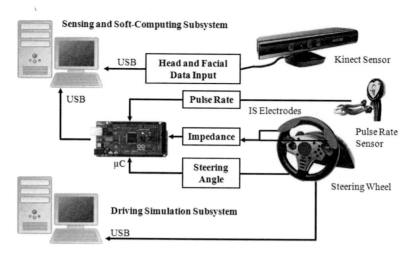

Figure 1. DeCaDrive: intelligent multi-sensor system for driver status monitoring with embedded impedance spectroscopy.

2 SYSTEM COMPONENTS AND MULTI-SENSOR INTERFACE

In the previous study, DeCaDrive system concept aimed at intelligent driver status monitoring has been realized through driving simulation, sensing and soft computing subsystems [7]. DeCaDrive system originated from a prototype based on single depth camera and afterwards has evolved to an intelligent multi-sensor system incorporating PC based driving simulation and embedded multi-sensor interface including depth camera (Kinect sensor), pulse rate sensor, blood oxygen saturation meter, steering angle sensor, tactile sensor and pressure sensor.

In this work DeCaDrive has been enhanced with embedded impedance spectroscopy to utilize (skin) impedance response on awake level for driver drowsiness detection. As illustrated in Fig. 1, data links of various embedded sensors on steering wheel are channelized to microcontroller based digital frontend so as to establish scalable adaptive multi-sensor interface. Depth camera, here Kinect sensor, as a key component of the sensing subsystem is connected to PC-based backend directly. The sensing and soft-computing subsystem is logically independent from the driving simulation subsystem despite that two subsystems can be unified in a single hardware environment. In the development phase of current approach two PCs are used for respective subsystems due to limitation of computation power. For future on-board device in vehicle embedded solutions will be employed.

3 EMBEDDED IMPEDANCE SPECTROSCOPY SENSOR

Bipolar electrode method is adopted here for impedance measurement, because not only tissue impedance but Galvanic Skin Response (GSR) of driver are under study. Both measures can be used as indication of psychological or physiological arousal in driving simulation context. In typical bioelectrical impedance analysis only the tissue impedance is of interest which is usually measured with tetrapolar electrode method [8]. Fig. 2 (a) illustrates bipolar electrode configuration on a single hand where the tissue impedance of palm (Z_{TUS}) is measured along with contact impedance at two electrode positions (Z_{k1} and Z_{k2}). Impedance measurement using bipolar electrode method can be applied on two hands as well. The overall measured impedance Z is interpreted as follows

$$Z = \frac{U}{I} = Z_{k1} + Z_{TUS} + Z_{k2} \qquad (1)$$

when contact impedance of two electrodes are the same, e.g., using one-hand measurement as Fig. 2 (a), where $Z_{k1} \approx Z_{k2} = Z_k$, Eq. (1) can be simplified to

$$Z = 2 \cdot Z_k + Z_{TUS} \qquad (2)$$

For non-intrusive impedance measurement two dry electrodes made of flexible copper tape (10 cm × 3 cm) were embedded on the steering wheel at standard "ten to two" position in the first prototype. As depicted in Fig. 2 (c) the measurement signal path is from one electrode over the body (including arm and thorax impedance) to the other electrode. One-hand and two-hand measurement configurations can be combined in a proper manner by considering different driving styles. With embedded impedance spectroscopy sensor the system is able to detect whether the steering wheel is handled with a single hand or two hands in addition to monitoring the impedance response. Such information can be used in the context of driving scenario to assess the driving behavior of individuals.

The dry electrodes being used to build embedded impedance spectroscopy sensor are having the following characteristics:

- They have good electrical conductance.
- They can be easily deformed to fit the steering wheel contour without causing discomfort for driver.
- The adaptation does not require additional parts or special treatment for the steering wheel.
- No skin preparation or additional electrolyte gel is required for the measurement.

However, the copper based dry electrodes are facing the limitations such as higher contact impedance in general, low biocompatibility and weak corrosion resistance of copper. For future long-term applications the copper electrodes shall be replaced by medically safe materials like stainless steel or aluminum. These metal electrodes could be attached to a precisely fitting milling groove in the steering wheel. Alternatively the steering wheel can be covered with conductive plastics or textile electrodes (with silver wires covered yarns or silver polyamide fabric) which allow flexible adaptation of the electrodes on the steering wheel contour despite very high impedance (Megaohm range) of textile electrodes. Another limitation of the current IS-sensor is less robust against movement artefacts like hand contact loss or high contact pressure to the electrodes which directly affect the measurement results. On the other hand such movement artefacts can be used to detect abnormal stress during driving, especially in conjunction with video cues and other sensor input, and hence may serve the assessment of driving performance.

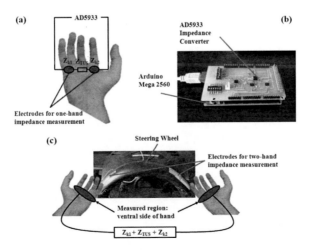

Figure 2. Embedded IS sensor configuration: one-hand measurement (a), two-hand measurement (c) and its analog front-end (b).

The embedded IS-system consists of self-designed IS-sensor on steering wheel, Arduino Mega 2560 microcontroller with shield board incorporating the impedance converter AD5933, Analog Front-End (AFE) to enable measurement of small impedance down to 100 Ω (see Fig. 2 (b)). Through the AFE useful adaptations have been introduced which are partially addressed in [9]. A highly accurate crystal-oscillator with 16 MHz and 25 ppm frequency stability was added along with two precise voltage references. The oscillator replaces the internal master clock with 16.667 MHz and 330 ppm so as to improve the stability of excitation voltage. In addition, a switch array of 12 ports were installed to realize a resistor network of high precision calibration and feedback resistors (tolerance ±0.05%) allowing a faster and easier calibration process. With AFE the DC level on the signal path remains constant and thus the effects of electrode polarization on sample impedance can be minimized.

4   FEATURE COMPUTATION AND SENSOR FUSION

The framework of presented intelligent multi-sensor system is reflected by its data processing flow as illustrated in Fig. 3. Diversified sensors in field and sophisticated algorithms make the system scalable and adaptive to different driving profiles and scenarios. Data sets of complementary sensors are synchronized on the same time base before being conveyed to feature computation components. Based on the outcome of feature computation selected data sets are fused on the feature level to construct input vectors for pattern classification so as to detect driver drowsiness. The classifier being used in this work is built upon Artificial Neural Network (ANN) or, more particularly, Multilayer Perceptrons (MLP) with supervised training procedure.

Various visual clues including head movement, eye gaze direction and ocular measures are computed from Kinect sensor data. The following features are extracted: mean value of head positions in 3D coordinate system of Kinect sensor within a measurement time frame; mean value of head orientation measures, i.e., pitch, yaw and roll; frequency domain analysis for head translation and rotation on three axes and its Euclidean norm based on FFT; head translation speed and head rotation speed; mean value of eyebrow positions relative to the respective eyes within a measurement time frame; mean value of eye blink frequency and blink duration within a measurement time frame.

The features computed from steering angle sensor are described as follows. Steering reversals being related to micro-corrections indicate the frequency of lateral motion changes (left-right or right-left) within gap size θ. Depending on θ two features are taken into evaluation (with $\theta = 1°$ and $\theta = 3°$ respectively). Steering-same-side represents the frequency of steering motion in the same direction above threshold ϑ which indicates lane changing or curve turning movements. Two features are computed based on $\vartheta = 12°$ and $\vartheta = 32°$ respectively. Mean and standard deviation of steering wheel positions within a measurement time frame are evaluated. The percentage of micro-corrections being taken to the overall steering

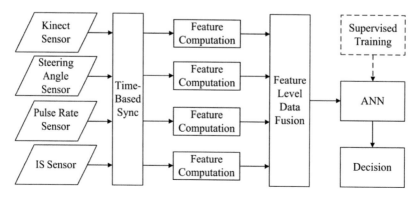

Figure 3.   Overview of the data processing flow of extended DeCaDrive system.

126

motion as well as the steering velocity are considered here. In addition, FFT based frequency domain analysis of steering statistics is performed to extract respective features. Parameters for feature computation here are depending on system calibration and steering wheel specifications such as steering wheel size, resolution of steering angle sensor. More details are addressed in [10], [11].

Based on embedded pulse rate sensor on steering wheel the low frequency to high frequency ratio of heart rate variation, or say pulse rate, in frequency domain is suggested as an indicator for drowsiness detection in [4]. With predefined Low Frequency band 0.04–0.15 Hz (LF) and High Frequency band 0.15–0.4 Hz (HF) the LF/HF ratio of pulse rate course within a measurement time frame is computed. The mean value of pulse rate is taken up in feature computation as well.

As discussed in Section 3 the self-designed IS-sensor is embedded in steering wheel and linked with impedance converter AD5933 which is capable of converting measured complex impedance to magnitude and phase angle output with frequency sweep procedure. Due to first cut implementation the equivalent circuit modeling for complex impedance by using, e.g., Cole-Cole plot technique is not performed here. Software tools such as ZView®ZV13 can be used for this purpose. A 3D *impedance spectrogram* is introduced instead to facilitate impedance response analysis in both time and frequency domains. As depicted in Fig. 4 the magnitude and phase angle of measured complex impedance are visualized in 330-second time frame with 10 times of frequency sweep from 10 to 100 kHz. A time frame covering one frequency sweep with specified frequency range is referred as evaluation window for IS-sensor data. In this work the impedance with magnitude of approx. 0.9 kΩ is observed at 30 kHz measurement frequency by applying metal electrodes to dry skin (see Fig. 5). Linear, exponential and polynomial regression analysis can be performed on IS-sensor data in evalu-

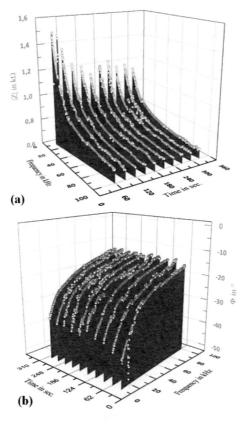

Figure 4.   3D impedance spectrogram: magnitude (a) and phase angle (b).

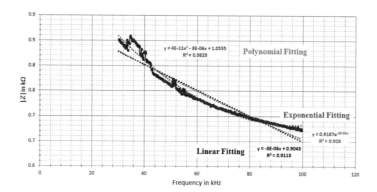

Figure 5. Linear, exponential and polynomial regression analysis for magnitude of complex impedance within one evaluation window of frequency sweep from 30 to 100 kHz.

ation window. Considering the goodness of fit of regression model (refer to R-squared value in Fig. 5) IS-sensor data including magnitude and phase angle is modeled as second-order polynomial in each evaluation window for feature computation. The IS-sensor features being extracted in the first cut implementation are summarized as follows:

- mean values of magnitude and phase angle
- standard deviations of magnitude and phase angle
- coefficients of quadratic polynomial fit to magnitude
- coefficients of quadratic polynomial fit to phase angle.

In order to synchronize IS-sensor with other embedded sensors in system the time base and data evaluation time frame being used during system runtime have been streamlined by referring to Kinect sensor time base and taking the least common multiple of respective sensor evaluation time frames.

Due to different driving styles, physiological conditions, etc., the dynamic range and variation of sensor data differ significantly among individual test subjects. In order to consolidate sensor features of respective test subjects feature vectors are normalized as per Eq. 3,

$$\vec{D}_{k,T,norm} = \frac{\vec{D}_{k,T,orig} - \mu_{k,T}}{\sigma_{k,T}} \tag{3}$$

where $\vec{D}_{k,T,norm}$ is the normalized feature vector of test subject T for specific sensor feature k, $\vec{D}_{k,T,orig}$ is the corresponding original feature vector, $\mu_{k,T}$ and $\sigma_{k,T}$ are mean value and standard deviation of $\vec{D}_{k,T,orig}$ respectively [10].

5 EXPERIMENTAL RESULTS

To investigate the potential correlation between driver drowsiness level and impedance response a set of experiments with five test subjects has been carried out in a simulated driving environment with room temperature (20 to 25 degrees Celcius). All participants are male with average age of 28 (±3) years old. One participant is not in possession of driving license while the other four drive regularly and have 7 to 12 years of driving experience.

The participants were instructed in advance to follow their normal daily routine and to avoid taking any stimulating substances (coffee, caffeine, etc.) before the experiment. The dry electrodes on wheel were finely prepared each time before a participant started with driving simulation. To increase occurrence of drowsiness a monotonous driving scenario (highway with low traffic density at daytime) was chosen for the experiment. The duration of driving session for each participant was set to 60 minutes. To minimize driving style associated influence all participants performed two-hand driving with proper skin-electrode contact area. In early phase of the experiment (2 to 5 minutes from the beginning) the measured impedance was biased due to temporal changes of dry electrode-skin interface (electrolyte diffusion process). As a result IS-sensor data of early phase was excluded from data evaluation process. Despite that sensor data evaluation can be performed during system runtime all the sensor measurements of driving simulation were time-based synchronized and recorded for more thorough offline analysis.

In the current system driver drowsiness detection is modeled as a three-class pattern classification problem. Therefore participants had rated their subjective sleepiness every 10 minutes based on predefined three-class scale, i.e., level 1 for alert, level 2 for drowsy and level 3 for deep drowsy. These subjective self-rating scores were used to establish ground truth. Due to advantage of learning complex, nonlinear, high-dimensional patterns a multilayer feedforward neural network is trained in a supervised manner for classification purpose. Two learning algorithms have been evaluated in this work, i.e., Scaled Conjugate Gradient (SCG) algorithm [13] and Levenberg-Marquardt (LM) algorithm [14]. The classification results were carried out by performing 10-fold cross-validation process. Table 1 and Table 2 give a comparison between two learning algorithms in terms of confusion matrix of classification. The classification accuracy (ACC) with dependency on the number of hidden neurons are illustrated in Fig. 6. With 40 hidden neurons the classifier based on LM algorithm achieved the ACC result of 99.6% with high performance and modest memory consumption in our experiments. The comparison among different feature sets are summarized in Fig. 7 where 8 SFS indicates a subset of 8 features being selected from the total feature set by applying Sequential Feature Selection algorithm (SFS) in the previous study [7]. These features are low

Table 1. Confusion matrix of classification results based on Scaled Conjugate Gradient (SCG) Algorithm.

SCG 80	Target class (GT)			
	I	II	III	Σ
Output Class (PR)				
I	339	22	10	91.4%
	7.1%	0.5%	0.2%	8.6%
II	147	3226	298	87.9%
	3.1%	67.7%	6.3%	12.1%
III	10	101	641	84.7%
	0.2%	2.1%	12.9%	15.3%
Σ	68.3%	96.3%	66.6%	87.7%
	31.7%	3.7%	33.4%	12.3%

Table 2. Confusion matrix of classification results based on Levenberg-Marquardt (LM) Algorithm.

LM 40	Target class (GT)			
	I	II	III	Σ
Output Class (PR)				
I	449	2	0	99.6%
	10.4%	0.0%	0.0%	0.4%
II	2	3338	4	99.8%
	0.0%	70.0%	0.1%	0.2%
III	0	9	918	99.0%
	0.0%	0.2%	19.3%	1.0%
Σ	99.6%	99.7%	99.6%	99.6%
	0.4%	0.3%	0.4%	0.4%

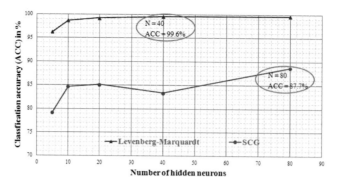

Figure 6.   Comparison of classification accuracy between SCG and LM algorithms.

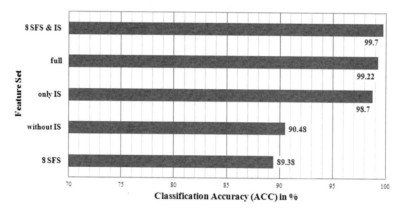

Figure 7.   Comparison of classification accuracy for different feature sets.

steering percentage, head position in 3D coordinate system of Kinect sensor, mean eyebrow position, mean blink frequency, pulse rate LF/HF ratio and mean pulse rate. Feature set *only* IS indicates IS-sensor features being addressed in Section 4 only. Feature set *without* IS means the total feature set excluding IS-sensor features, while *full* represents the complete feature set. The combination of *8 SFS and IS* yielded the best result in the experiment.

6   CONCLUSION

Active safety related driver assistance systems have paved the way to mainstream automotive applications. Despite the trend that future automotive electronics advances towards fully autonomous driving, such system as presented in this work belongs to the fundamental components of human-vehicle-interaction in active safety context, hence can contribute to human centered safe mobility.

In our first cut implementation impedance spectroscopy has been investigated with a performance limited impedance converter chip for a potential mass market application. The self-designed embedded IS-sensor has been integrated in DeCaDrive system to improve driver drowsiness detection and to facilitate general driver status monitoring. The experimental results in this work support the hypothesis and confirm the finding that impedance response analysis can improve the effectiveness of driver drowsiness detection.

In addition to MLP based classifier the presented approach should be validated with other advanced classification techniques such as Support Vector Machine (SVM). The robustness should be further examined with more statistics and with data from real driving scenarios. Inspired by study in [15] driving impairment such as inebriation should be investigated by utilizing bioelectrical impedance with tetrapolar electrode method. Correlation study of bioelectrical impedance and human emotional state can be carried out thereafter.

We work on amelioration in this explorative work, including better electrodes, electronics, sensor context to overcome deficiencies due to impedance converter, electrodes, hand operation context such as pressure and sweat. More sophisticated instrumentation or (CMOS) circuit integration should be considered. The scalable adaptive multi-sensor interface and PC-based soft-computing subsystem can be harmonized towards System-on-Chip for high-performance and low-power solution in the future.

REFERENCES

[1] N.S. Foundation. (2009). Facts and Stats. (visited on 04/13/2014), [Online]. Available: http://drowsydriving.org/about/facts-andstats.

[2] C. Evers and K. Auerbach, "Verhaltensbezogene Ursachen schwerer Lkw-Unfälle," BASt-Bericht M 174, Tech. Rep., 2005.

[3] C. Zocchi, A. Giusti, A.Rovetta, *et al.*, "Biosensors for microsleeps detection during drive simulations," *Biodevices 2008 International Conference Proceedings*, 2008.

[4] X. Yu, "Real-time Nonintrusive Detection of Driver Drowsiness," 2009.

[5] M. Walter, B. Eilebrecht, T. Wartzek, *et al.*, "The smart car seat: personalized monitoring of vital signs in automotive applications," *Personal and Ubiquitous Computing*, vol. 15, no. 7, pp. 707–715, 2011.

[6] V. Savchenko, "Monitoring of an operator's vigilance level by skin resistance response," *Control Engineering Practice*, vol. 4, no. 1, pp. 67–72, 1996.

[7] L. Li, K. Werber, C.F. Calvillo, *et al.*, "Multi-Sensor Soft-Computing System for Driver Drowsiness Detection," in *Soft Computing in Industrial Applications*, Springer, 2014, pp. 129–140.

[8] A. Ivorra, "Bioimpedance Monitoring for physicians: an overview," *Centre Nacional de Microelectrònica Biomedical Applications Group*, 2003.

[9] A. Devices, *Datasheet ad5933-rev. e*, USA, 2013.

[10] K. Werber, "Untersuchung von Fahrerassistenzsystemen zur Fahrer-, Zustands- und Absichtserkennung mit Multisensorik," Master's thesis, TU Kaiserlautern, 2012.

[11] P.J. Sherman, M. Elling, and M. Brekke, "The Potential of Steering Wheel Information to Detect Driver Drowsiness and Associated Lane Departure," Tech. Rep., 1996.

[12] (2013). Zview for Windows. visited on 10/29/2014, [Online]. Available: http://www.scribner.com.

[13] M.F. Møller, "A Scaled Conjugate Gradient Algorithm for Fast Supervised Learning," *Neural networks*, vol. 6, no. 4, pp. 525–533, 1993.

[14] J.J. Moré, "The Levenberg-Marquardt algorithm: implementation and theory," in *Numerical analysis*, Springer, 1978, pp. 105–116.

[15] M. Ulbrich, M. Czaplik, A. Pohl, *et al.*, "Estimation of Blood Alcohol Content with Bioimpedance Spectroscopy," *Abstract Book, Int. Workshop on Impedance Spectroscopy IWIS 2013*, pp. 52–53, 2013.

Lecture Notes on Impedance Spectroscopy, Volume 5 – Kanoun (Ed.)
© 2015 Taylor & Francis Group, London, ISBN 978-1-138-02754-1

Author index

ADVANCES IN ENERGY AND ENVIRONMENT RESEARCH